ENVIRONMENTAL INFLUENCES ON FERTILITY, PREGNANCY, AND DEVELOPMENT
Strategies for Measurement and Evaluation

ENVIRONMENTAL INFLUENCES ON FERTILITY, PREGNANCY, AND DEVELOPMENT

Strategies for Measurement and Evaluation

Proceedings of a meeting held in Cincinnati, Ohio,
May 24 and 25, 1982

Editors

MARVIN S. LEGATOR, PhD
Division of Genetic Toxicology
University of Texas Medical Branch
Galveston, Texas

MICHAEL J. ROSENBERG, MD, MPH
Family Health International
Research Triangle Park, North Carolina

HAROLD ZENICK, PhD
Department of Environmental Health
University of Cincinnati Medical Center
Cincinnati, Ohio

---

also published as Volume 4, Number 1, 1984
of the journal
Teratogenesis, Carcinogenesis and Mutagenesis

ALAN R. LISS, INC. • NEW YORK

These papers are being printed both as an issue of Teratogenesis, Carcinogenesis, and Mutagenesis, Volume 4, Number 1, 1984 with Dr. Marvin S. Legator, Dr. Michael S. Rosenberg, and Dr. Harold Zenick as Editors, and as a separate volume, Environmental Influences on Fertility, Pregnancy, and Development: Strategies for Measurement and Evaluation, edited by Dr. Marvin S. Legator, Dr. Michael S. Rosenberg, and Dr. Harold Zenick.

Address all Inquiries to the Publisher
Alan R. Liss, Inc., 150 Fifth Avenue, New York, NY 10011

Copyright © 1984 Alan R. Liss, Inc.

Printed in the United States of America.

Under the conditions stated below the owner of copyright for this book hereby grants permission to users to make photocopy reproductions of any part or all of its contents for personal or internal organizational use, or for personal or internal use of specific clients. This consent is given on the condition that the copier pay the stated per-copy fee through the Copyright Clearance Center, Incorporated, 21 Congress Street, Salem, MA 01970, as listed in the most current issue of "Permissions to Photocopy" (Publisher's Fee List, distributed by CCC, Inc.), for copying beyond that permitted by sections 107 or 108 of the US Copyright Law. This consent does not extend to other kinds of copying, such as copying for general distribution, for advertising or promotional purposes, for creating new collective works, or for resale.

Library of Congress Cataloging in Publication Data

Main entry under title:

Environmental influences on fertility, pregnancy, and development.

    Includes bibliographies and index.
    1. Environmentally induced diseases—Congresses.
2. Infertility —Etiology—Congresses. 3. Pregnancy, Complications of—Congresses. 4. Abnormalities, Human—Etiology—Congresses. I. Legator, Marvin S., 1926–
II. Rosenberg, Michael. III. Zenick, Harold. [DNLM:
1. Environmental exposure—Congresses. 2. Evaluation studies—Congresses. 3. Fertility—Drug effects—Congresses. 4. Mutagenicity tests—Congresses.
5. Probability—Congresses. 6. Reproduction—Drug effects—Congresses. 7. Spermatozoa—Drug effects—Congresses.
8. Toxicology—Congresses.   WP 565 E61 1982]
RB 152.E57   1984   618.1′78071   83-23869
ISBN 0-8451-0230-3

Contents

Contributors	vii
Preface Marvin S. Legator, Michael J. Rosenberg, and Harold Zenick	ix
A Framework for Reproductive Risk Assessment and Surveillance Gilbert S. Omenn	1
The Role of Surveillance in Monitoring Reproductive Health Michael J. Rosenberg and William E. Halperin	15
Field Studies: Lessons Learned Donald M. Whorton	25
The Effect of Common Exposures on Reproductive Outcomes Carol J.R. Hogue	45
Fertility as a Measurement in Reproductive Toxicology Anthony R. Scialli and Sergio E. Fabro	59
Laboratory Tests for Human Male Reproductive Risk Assessment James W. Overstreet	67
An Evaluation of Sperm Tests as Indicators of Germ-Cell Damage in Men Exposed to Chemical or Physical Agents Andrew J. Wyrobek, George Watchmaker, and Laurie Gordon	83
Evaluating Male Reproductive Toxicity in Rodents: A New Animal Model Harold Zenick, K. Blackburn, E. Hope, D. Oudiz, and H. Goeden	109
The Mouse as a Model System for Mutation Testing and Evaluation of Risk in Mammals Susan E. Lewis	129
Childhood Tumors and Parental Occupational Exposures John M. Peters and Susan Preston-Martin	137
Male-Transmitted Developmental and Neurobehavioral Deficits Perrie M. Adams, O. Shabrawy, and Marvin S. Legator	149
Index	171

Contributors

Perrie M. Adams, Department of Psychiatry and Behavioral Science; and Department of Pharmacology and Toxicology, The University of Texas Medical Branch, Galveston, TX 77550 [149]

K. Blackburn, Department of Environmental Health, University of Cincinnati, Cincinnati, OH 45267 [109]

Sergio E. Fabro, Reproductive Toxicology Center, Columbia Hospital for Women, Washington, DC 20037 [59]

H. Goeden, Department of Environmental Health, University of Cincinnati, Cincinnati, OH 45267 [109]

Laurie Gordon, Lawrence Livermore National Laboratory, Biomedical Sciences Division, University of California, Livermore, CA 94550 [83]

William E. Halperin, National Institute for Occupational Safety and Health, Centers for Disease Control, Cincinnati, OH 45226 [15]

Carol J.R. Hogue, Pregnancy Epidemiology Branch, Centers for Disease Control, Center for Health Promotion and Education, Division of Reproductive Health, Atlanta, GA 30333 [45]

E. Hope, Department of Environmental Health, University of Cincinnati, Cincinnati, OH 45267 [109]

Marvin S. Legator, Department of Preventive Medicine and Community Health, The University of Texas Medical Branch, Galveston, TX 77550 [149]

Susan E. Lewis, Research Triangle Institute, Research Triangle Park, NC 27709 [129]

Gilbert S. Omenn, Department of Medical Genetics, University of Washington, Seattle, WA 98195 [1]

D. Oudiz, Department of Environmental Health, University of Cincinnati, Cincinnati, OH 45267 [109]

James W. Overstreet, School of Medicine, University of California, Davis, Davis, CA 95616 [67]

John M. Peters, University of Southern California, School of Medicine, Department of Family and Preventive Medicine, Los Angeles, CA 90033 [137]

Susan Preston-Martin, University of Southern California, School of Medicine, Department of Family and Preventive Medicine, Los Angeles, CA 90033 [137]

Michael J. Rosenberg, National Institute for Occupational Safety and Health, Centers for Disease Control, Cincinnati, OH 45226 [15]

Anthony R. Scialli, Reproductive Toxicology Center, Columbia Hospital for Women, Washington, DC 20037 [59]

O. Shabrawy, Department of Preventive Medicine and Community Health, The University of Texas Medical Branch, Galveston, TX 77550 [149]

George Watchmaker, Lawrence Livermore National Laboratory, Biomedical Sciences Division, University of California, Livermore, CA 94550 [83]

Donald M. Whorton, Environmental Health Associates, Inc., Berkeley, CA 94704 [25]

Andrew J. Wyrobek, Lawrence Livermore National Laboratory, Biomedical Sciences Division, University of California, Livermore, CA 94550 [83]

H. Zenick, Department of Environmental Health, University of Cincinnati, Cincinnati, OH 45267 [109]

The number in brackets is the opening page number of the contributor's article.

Preface

In this special issue several timely articles appear having to do with environmental influences on fertility, pregnancy, and reproductive outcomes. These papers were initially presented in a workshop on May 24 and 25, 1982 and the various studies have been updated for presentation in this volume. Participants attempted to address research priorities, the utility of various research strategies, methodological and design considerations in the selection of end points to be evaluated, and the contribution of animal data to human studies. The information emerging from these proceedings will hopefully help better direct research in this rapidly growing area of concern.

In an overview of the papers in this issue it is apparent that we are discussing a spectrum of chemically induced adverse health outcomes including prezygotic effects, effects during in utero exposure, as well as genetic damage induced during spermatogenesis and detected in the F_1 generation.

- The paper by Zenick and his co-workers deals with teratogenic studies in rodents.
- The paper by Lewis reviews various approaches for evaluating genetic damage in the mouse including identifying genetic mutations by scoring mice for various enzyme activities.
- The paper of Adams and his colleague describes the induction of behavioral anomalies in rats following chemical exposure during spermatogenesis.
- The sperm morphology studies of Wyrobek presents a technique for monitoring adverse effects in both animals and man.
- The studies of Omenn, Rosenberg and Halperin, Whorton, Hogue, and Scialli and Fabro summarize the status of human monitoring studies for reproductive outcomes, and present examples of what has been learned from our past experiences with chemicals such as Dibromochloropropane.
- The work reported by Peters and Preston-Martin indicates childhood tumors after paternal exposure.

We believe this is a timely issue which discusses an important area having to do with effects of mutagenic and teratogenic agents.

Marvin S. Legator, Ph.D.
Michael J. Rosenberg, M.D., M.P.H.*
Harold Zenick, Ph.D.

*Dr. Rosenberg was previously with Reproductive Hazards Assessment Branch, NIOSH, Cincinnati, Ohio.

A FRAMEWORK FOR REPRODUCTIVE RISK ASSESSMENT AND SURVEILLANCE

Gilbert S. Omenn, M.D., Ph.D.
Professor and Chairman, Environmental Health
Professor of Medicine, Medical Genetics
University of Washington
Seattle, WA 98195

As others have, I should explain the origins of my involvement in this important meeting. My interests have several sources. First, as a physician, a clinician, I see patients with these problems. I am interested not only in the clinical complexity, but also in the mechanisms and population base from which affected patients appear. Second, as a biochemical geneticist, I am interested in individual differences or population subgroup differences in susceptibility to particular environmental hazards. Mechanisms of adverse effects are surely many, and potential preventive interventions and treatments may have to be specific for different classes of mechanisms. Third, I had the extraordinary experience from 1977 to 1980 of serving as deputy science adviser to President Carter under Dr. Frank Press in the White House Office of Science and Technology Policy (OSTP). During that period of time, it became clear that strengthening the science base for regulation and for public education about environmental and occupational hazards was an exceedingly important role for us. I worked with the Interagency Regulatory Liaison Group and represented OSTP on the Regulatory Analysis Review Group, where we grappled with important regulations in such areas as air pollution, food additives, occupational carcinogens, and waste dumps. During the final year of the Carter administration, I served as an Associate Director of the Office of Management & Budget, where questions such as those raised by Michael Rosenberg and other speakers about cost-effectiveness, efficiency, trade-offs, and investment of public resources are the mainstream activity. Fourth, I have been involved during the past year with the Hastings Center Institute of Society, Ethics, and Life Sciences and

© **1984 Alan R. Liss, Inc.**

the Congressional Office of Technology Assessment in studies of scientific, ethical, and legal aspects of screening in the workplace. Finally, I am back now at the University of Washington, chairing the Department of Environmental Health and Occupational Medicine, in which we have a NIOSH-sponsored Educational Resource Center and substantial interests in potential reproductive hazards.

We face an important challenge in the widespread misperception among the public and even among scientists that practically every chemical will prove to be carcinogenic or mutagenic or otherwise hazardous if tested long enough and at high enough doses. Cartoonists and lyricists have had a field day with our predicament and the frequent news that the foods we eat, the air we breathe, and the water we drink are all hazardous to our health. For example, the *National Journal* (a fine political news magazine in Washington, D.C.) ringed the pages of an article about the 1978 debate about regulation of saccharin with pictures of dead rats, all of them tumorous. Under each was written the presumed cause, ranging from blueberry waffles and char-broiled hamburgers, to air, water, coffee, saccharin, and aphrodisiacs (the latter dead rat bearing a smile)! A lyricist lightened the program of a 1981 Food Safety Conference with these illustrated verses:

> "'Twas the night before Christmas and all through the house
> Not a creature was stirring, not even a mouse.
> With 100 adults and a tumbler of scotch
> One had smoked cigarettes as he kept watch.
> One had used saccharin to sweeten her sherry,
> Some swallowed red dye from a maraschino cherry.
> A few crumbs of bacon were found where they'd been dropped;
> Nearby was evidence of spilled diet pop.
> Beer was drunk heartily and all went to bed;
> By morning the Surgeon General had pronounced them all dead."

A FRAMEWORK FOR DECISION-MAKING

Table 1 outlines a framework for the scientific and regulatory phases of decisions by governmental agencies or by manufacturers about potentially hazardous chemicals. My staff and I in the Office of Science & Technology Policy developed this framework with considerable input and review from colleagues throughout the government, academia, industry, and environmental groups (Calkins, Dixon, Gerber, Zarin, Omenn 1980). The tasks of identification, characterization, and control of such chemical hazards are divided into two stages: first, the scientific work and scientific review, and, second, the regulatory process which must lead to decisions about appropriate actions to take.

TABLE 1

Potentially Hazardous Chemicals

FRAMEWORK FOR DECISION-MAKING IN THE UNITED STATES

1.	IDENTIFICATION	EPIDEMIOLOGY ANIMAL TESTS IN VITRO SCREENING
2.	CHARACTERIZATION	RELATIVE POTENCY EXPOSURES RANGE OF SUSCEPTIBILITY
3.	CONTROL	INFORMATION REGULATION SUBSTITUTION

Identification

There are many endpoints for health effects studies of chemicals and other environmental agents (Table 2).

TABLE 2

MAJOR HEALTH END-POINTS

Carcinogenicity
Mutations
Teratogenicity: birth defects
Altered Reproductive Function
Neuro-behavioral toxicity
Other specific organ system effects

Carcinogenicity has been tested most. Increasing emphasis, however, is being given to mutations, birth defects, and altered reproductive function. Many chemicals produce a

variety of health effects, making clear the importance of the
first step in our Framework, the description of the evidence
and qualitative judgment that specific health effects may,
indeed, be associated causally with exposures to a particular
chemical.

The identification step requires a yes-no decision.
However, we must recognize that "positive" and "negative"
results are not qualitatively different. Laboratory and epidemiological studies make observations, compare them with
expectations, and draw conclusions with all the statistical
conventions, extrapolation assumptions, and uncertainties
involved. A few more or a few fewer observed abnormalities
may shift a particular chemical into the positive or the
negative category, respectively. The investigational
approaches are listed in Table 1 and have been discussed by
others in the Conference. It is important to note that for
many of the effects we are examining, especially the reproductive effects, the proper study of mankind may, indeed, be
men and women. The task of extrapolating effects on various
reproductive processes across species, as complicated as
they are in any one species, is a difficult challenge. The
animal tests, however, remain extremely important for chemicals and other substances not yet introduced into wide use
with human exposures. Many companies have avoided further
development of chemicals which on initial screening proved
to be toxic in various tests. Animal studies also are
extremely valuable for investigations of mechanisms, getting
inside that "black box" of statistical associations. Finally,
there are numerous possibilities for in vitro screening tests
in the reproductive arena, just as for mutagenic and carcinogenic effects. Participants in this Conference have been
leaders in this work, which is well summarized in a recent
book published by the March of Dimes (Bloom 1981).

The difficulties of determining what should be considered an adverse health effect can be illustrated with the
1979 revision of the National Ambient Air Quality Standard
for photochemical oxidants measured as ozone (Environmental
Protection Agency 1978). Pulmonary symptoms were considered
the relevant health effect. Most lay people would identify
eye irritation as the most common and most troubling effect
of smog. Yet, for a variety of reasons, the EPA has treated
eye irritation as a "welfare" effect, rather than a "health"
effect. Another example arose in the standard setting for

airborne lead (Environmental Protection Agency 1978). Should the adverse health effect to be prevented be nervous system damage, or anemia, or increases in blood lead level? Greatest emphasis, in fact, was placed on zinc protoporphyrin levels in red blood cells, as biochemical indicators of lead exposure and of altered heme biosynthesis. No adverse physiological effects have been associated with modest increases of protoporphyrin. We can expect similar dilemmas in the reproductive area.

Characterization

The identification of the potential of a chemical to have adverse effects is just the beginning. There has been a lot of unproductive debate about whether or not quantitative risk assessments, with all the uncertainties involved, should even be done. My view is that we must do quantitative risk assessments when the data are good enough, but we must never lose sight of the qualitative aspects of the nature of the health effects, the strength of the evidence, the kinds and levels of exposures, and the differential susceptibility within the population. The regulatory officials have tremendous discretion in evaluating potential hazards, deciding which ones merit priority attention, and promulgating appropriate actions. An attempt to perform quantitative risk assessment adds discipline to these discretionary decisions, forcing closer attention to the quantitative details of the experiments, the frequency of adverse effects, the dose/response relationships, the exposures to humans, and the relationships between experimental or occupational exposures and the exposure levels of the general population which may be the target for regulatory action.

I urge the use of the term "characterization", instead of "quantitation", of risk, because both qualitative and quantitative information must be brought together in the assessment of risk to humans. Three main elements are involved (Table 1): potency, exposures, and susceptibility.

Estimates of <u>relative potency</u> are feasible from dose/response relationships within any single test system. As testing protocols are standardized and utilized more widely, there will be more and more basis for comparison of data on different chemicals from the same test systems. Results may be expressed as dose required to cause 50 percent of animals to exhibit a particular effect, i.e., ED_{50}, with adjustments according to life table analyses. Alternatively, potency may be expressed in terms of probability of effects per unit dose of compounds; EPA uses this approach, with $\mu g/m^3$ as the unit dose, for analyses by the Health Effects Assessment Group.

Two kinds of extrapolation are necessary to estimate potency for humans: from high dose to low dose in animals, and from low dose in animals to low dose in humans (see Office of Technology Assessment 1981). Estimated effects at doses well below the observable range are highly sensitive to the choice of model for extrapolation, whereas most of the models fit the data in the observed relatively high range about equally well. The linearized multistage model developed by Crump et al. (1976) has been adopted recently by EPA and has the advantage of being compatible with current thinking about the multiple stages of initiation and progression of tumors. It should be appropriate also for the complex points of action for reproductive toxicity. The linearized multistage model has the further advantage of utilizing all the data points for those experiments which have more than one dose with interpretable results. It is critical that a particular extrapolation model and a particular scaling factor (from rodents to humans) be used consistently, in order to permit ranking of relative potency and in order to eliminate this unnecessary source of variation and uncertainty in analyses.

When different results are obtained from different species, it is important to try to understand whether one or another test species more closely resembles humans for the relevant underlying mechanisms. In the absence of firm evidence that the species with positive test results is strikingly different from humans in metabolism of the chemical or susceptibility to its effects and that the species with negative test results is similar, it is generally agreed that the "sensitive" species should be used as the basis for extrapolation to humans. Laboratory animals are highly inbred, thus showing less intra-species variability than humans. Negative results, especially negative results in the same species, should be taken into account in the qualitative assessment of the weight of evidence.

Next, exposures to humans must be characterized carefully. In general, information about exposures is woefully inadequate and neglected. It is essential to determine the sources, prevalence, levels, and time-course of exposures. Cooperation from industry is often crucial to determine how many people are exposed in the various classes of exposures. For reproductive effects, it is necessary to show that the agent actually penetrates to the targets for action in the reproductive organs.

Characterization of risk also should examine differences in susceptibility in the human population. Age, sex, genetic predispositions, exposure to other chemicals, and important lifestyle factors such as smoking deserve attention. Curiously, this factor has gained little attention from the research community, but a lot of attention from the regulators. The laws governing the regulatory programs in the U.S. for control of air pollution and protection of worker safety and health explicitly require EPA and OSHA, respectively, to set standards so as to protect the most susceptible groups or individuals in the population (Omenn 1979; Friedman 1979). Under section 109 of the Clean Air Act, EPA focused on the 3-5 percent of the population with asthma, bronchitis, or emphysema in establishing the air quality standard for ozone, aiming in one aspect of the analysis to protect 99 percent of this subgroup. A similar process in the air lead standard-setting was directed at protecting 99.5 percent of exposed children. It is a matter of judgment and administrative discretion whether 99.5 or 99 or 95 or some other percentage of the population subgroup should be the goal for the protective standard. These examples serve to illustrate that other important health effects besides carcinogenicity share the problems of extrapolation to low doses and of individual variation in susceptibility, making determination of a no-effect threshold impossible. There are many clues to underlying genetic predispositions to particular classes of chemicals, as summarized in Table 3 and discussed elsewhere (Omenn 1982). Previous occupational exposures is an important class of predisposing conditions, as was reported recently for three former workers with silicosis who developed life-threatening infections with Acinetobacter, usually considered to be innocuous organisms (Cordes, Brink, Checko, Lentnik, Lyons, Hayes, Wu, Tharr, Fraser, 1981).

TABLE 3

GENETIC VARIATION IN HUMAN SUSCEPTIBILITY TO ENVIRONMENTAL AGENTS

DRUGS

    Metabolism: N-Acetyl Transferase
                    Plasma Pseudocholinesterase
                    C-oxidation
    Sensitivity: Glucose-6-phosphate dehydrogenase deficiency
                    Methemoglobin reductase deficiency

INHALED POLLUTANTS/PESTICIDES

    Alpha-1 anti-trypsin deficiency
    Aryl hydrocarbon hydroxylase inducibility
    Metabolic conversion of nicotine
    Plasma paraoxonase activity

FOODS

    Lactose intolerance   Fava beans
    Wheat gluten/celiac disease   Goitrogens
    Saturated fats/atherosclerosis Catecholamines

FOOD ADDITIVES AND STIMULANTS

    Iron   Ethanol   Caffeine

PHYSICAL AGENTS

    Cold weather, motion sickness, metal poisoning, color vision, UV radiation, X-radiation

INFECTIOUS AGENTS/AUTOIMMUNE DISORDERS

    Malaria: Duffy blood group; Sickle hemoglobin, thalassemia, G6PD deficiency
    Predispositions due to impaired antibody production or cellular immunity
    Predispositions associated with histocompatibility phenotypes

---

From Omenn, Motulsky 1978

Regulatory Decisions

Examples of the uses of quantitative risk assessments are given in Table 4, drawing upon the experiences of various program offices at EPA and of the FDA. Their statutes require balancing of risks and benefits in many cases, especially for pesticides, toxic substances, drinking water, and food contamination. In certain other cases the agencies have decided that target levels of risk, that is tolerable remaining risks, must be identified. If the effects associated with the level of contaminant or other pollutant is below the target level, no action or less action will be taken.

TABLE 4

APPLICATIONS OF QUANTITATIVE RISK ASSESSMENT

1. Balancing Risks and Benefits

    FIFRA, TSCA, Drinking Water Act

2. Setting Target Levels of Risk

    FDA for food contaminants

    EPA, water pollution control guidelines

3. Setting Priorities for Program Activities

    EPA, Clean Air Act section 112

4. Estimating Residual Risks

    Assessing effects of best available technologies

Implementation of the Clean Air Act section 112 has been bedevilled by the requirement that hazardous air pollutants be regulated to protect even the most susceptible and provide an "ample margin of safety". For several years now, no airborne carcinogens have been regulated, despite a long list of candidate chemicals. Under Douglas Costle, EPA decided that it was necessary to challenge the Congressional language by declaring that EPA should accept some remaining risk from these chemicals in order to have a rational approach to reducing risk without closing down essential production. Finally, in cases where best available technologies have been applied, as is commonly the case with engineering controls, it is useful to estimate the risk after actions have been, or will have been, implemented to see whether sufficient protection has been achieved or whether further action would be a high priority.

RISK ASSESSMENT FOR GERM CELL MUTAGENIC EFFECTS

The EPA published proposed guidelines in 1980 for germ cell mutagenicity risk assessment (Environmental Protection Agency 1980). The direct demonstration of increased frequency of germ cell mutations in humans is extremely difficult. In the exhaustive studies of children of persons subjected to atomic irradiation in Japan, chromosomal aberrations, protein variants, sex ratio, and birth defects frequencies show little or no difference as a result of the radiation history (Schull, Otaka, Neel 1981). *In vitro* studies of human cells, employing selective media for detection of HGPRT mutants among cultured lymphocytes or fluorescent cell sorter techniques for immunochemical detection of variant hemoglobins in red blood cells, offer potentially powerful tools for estimation of somatic mutation rates (Bloom 1981). Analogous specific locus mutations may be identified in sperm.

The EPA risk assessment guidelines adopt a weight-of-evidence approach for available animal testing systems. A positive response in the mouse specific locus test is considered sufficient to regard a chemical as a potential mutagen for humans. Alternatively, positive responses in any two different point mutation test systems plus evidence for the presence of the test substance and/or its metabolites in mammalian gonadal organs will suffice. High weight was given also to the heritable translocation test and to the X-chromosome-loss test in mice, chromosomal alterations that are transmitted to progeny. Alternatively, positive results in two *in vivo* somatic cell cytogenetic tests or one *in vivo* and one *in vitro* test plus evidence of reaching the gonadal organs will be considered strong evidence that the compound is a potential mutagen for humans. The Guidelines explicitly give weight to consistency of results in different test systems and different species and to the quality of the test protocols.

Evidence against mutagenic potential can come both from lack of positive results in the appropriate test system and from evidence that the chemical does not reach or affect gonadal organs--based upon radioactive tracer studies, search for alkylated DNA or other cellular macromolecules, unscheduled DNA synthesis, sister chromatid exchange, or chromosomal aberrations in gonadal tissues.

Quantitative risk assessment for mutagenicity depends upon the qualitative results, the relative potency, and the potential human exposures. EPA proposed to use linear or linearized multistage no-threshold models for point mutations, but a multi-hit model for chromosomal effects.

RISK ASSESSMENT FOR REPRODUCTIVE ADVERSE OUTCOMES

As in the case of carcinogenesis or mutagenesis, recognition of adverse effects of chemicals upon reproductive functions of men and women and upon the fetus and offspring depends upon information from three major sources:

(1) alert observations by astute clinicians or workers or others
(2) systematic toxicological testing in non-human systems
(3) epidemiological studies and surveillance among human populations.

Of these three approaches, the observational report has been the most productive to date. Classic examples are the high prevalence of cataracts in children born of mothers with rubella noted by Gregg in 1941; the association between thalidomide and limb-reduction defects; the fetal alcohol and fetal hydantoin syndromes; the causal relationship between maternal diethylstilbestol and clear cell adenocarcinoma of the vagina; and the efforts by workers in a plant manufacturing dibromochloropropane (DBCP) to gain an explanation for their infertility. On the other hand, some associations prove to be coincidental after detailed analysis, and with so many thousands of chemicals around us and so many reproductive problems already common in the "background", some more systematic screening and testing seems warranted.

Toxicological studies are advancing in this arena (Environmental Protection Agency 1982). Investigations of chemicals for estrogen-like, androgen-like, or non-steroidal toxicant effects on sperm number and motility, conception, pregnancy fetal development, lactation, and reproductive performance of the offspring are being performed and tests of neurobehavioral toxicity, oocyte toxicity, cellular receptor responses, and *in vitro* model systems are being developed. A good example of the latter is the rat embryo test system employed by Fantel, Shepard, Juchau, Faustman-Watts and their colleagues (Faustman-Watts et.al, 1982).
The crucial role of bioactivation of 2-acetylaminofluorene has been demonstrated in 10-day rat embryos *in vitro*. 2-AAF itself has no effect on development. However, incubation with a microsomal monooxygenase system activated this mutagen and carcinogen to become a potent teratogen, as well. Two metabolites, the N-hydroxy and N-acetoxy derivatives, each produced malformations, but of different types than the activated AAF. Numerous modifications of the basic experiment have permitted search for the specific toxic metabolites for particular phenotypes, investigation of the sensitivity of the embryo to timing of exposures, and dissociation of embryotoxic and teratogenic effects. Detailed study of instructive compounds for teratogenic, reproductive, and mutagenic effects as well as carcinogenic effects should help to advance our understanding of the underlying mechanisms.

A host of end-points exists for surveillance and epidemiological studies in humans. These end-points have been described and their frequencies in populations summarized in the March of Dimes book (ref. 2, p. 47). The statistical requirements for detecting a significant increase in frequency of any of these phenotypes depend upon the general population frequency. More nearly complete ascertainment increases the frequencies and also the statistical power. As Oakley has described at this meeting, the Centers for Disease Control maintains and utilizes two well-developed and complementary surveillance systems to assess birth defects; these are the Metropolitan Atlanta Congenital Defects Program, with 24,000 births per year, and the Birth Defects Monitoring Program, with 1 million births per year at some 1200 hospitals (see 2, p. 56). These databases have yielded trend-lines for frequencies of some 150 birth defects. Highlights of the findings over the decade from 1970 to 1981 are three-fold increases in ventricular septal defects and in patent ductus arteriosus and decreases nationally in neural tube closure defects despite an increase regionally in Appalachia. Fortunately, no new "thalidomides" have been encountered. However, case-control studies have been useful in identifying a (weak) association between diazepam and facial clefts, a confirmation of the fetal hydantoin syndrome, and reassurance about claimed teratogenic risks from airport noise, spray adhesives, vinyl chloride, Bendectin, and progesterones.

Other data banks are available that should be linked into reproductive surveillance efforts. For example, the National Center for Health Statistics conducts an annual health interview survey which reaches some 40,000 households (120,000 people) in 376 defined geographical districts all around the country. This source is being utilized together with air pollution monitoring data obtained from EPA in a micro-epidemiological study by Resources for the Future (Portney, personal communication) which correlates individuals' exposures to a host of air quality measurements with their health status over the two-week periods covered in the Health Interview Survey.

FUTURE DIRECTIONS

Support for work in reproductive toxicology and reproductive risk assessment is sorely needed and is responsive to significant fears among the public and significant concern among manufacturers about hazards to workers and consumers and resulting liability. The systematic development of risk assessment for teratogenic and reproductive effects should follow the paths already being laid in the characterization of carcinogenic and mutagenic effects, and should benefit from the current efforts to correlate mutagenic and carcinogenic activities of specific chemicals and to understand the basis for such manifestations of chemical toxicity.

Bloom AD (ed) (1981). Guidelines for studies of human populations exposed to mutagenic and reproductive hazards. March of Dimes Birth Defects Foundation. New York.

Calkins DR, Dixon RL, Gerber CR, Zarin D, Omenn GS (1980). Identification, characterization, and control of potential human carcinogens: a framework for Federal decision-making. J Natl Cancer Inst 64:169-175.

Cordes LW, Brink EW, Checko PJ, Lentnik A, Lyons RW, Hayes PS, Wu TC, Tharr DG, Fraser DW (1981). Cluster of acinetobacter pneumonia in foundry workers. Annals Int Med 95:688-693.

Crump KS, Hoel DG, Langley CH, Peto R (1976). Fundamental carcinogenic processes and their implications for low dose risk assessment. Cancer Research 36:2973-2979.

Environmental Protection Agency (1978). EPA proposed national ambient air quality standard for lead. Fed Reg 42:63076, 1977. Final standard Fed Reg 43:46246.

Environmental Protection Agency (1978). Air quality criteria for ozone and other photochemical oxidants. EPA-600/8-78-004, April 1978. EPA Report on ozone air quality standard to regulatory analysis review group, 21 December 1978.

Environmental Protection Agency (1980). Mutagenicity risk assessments: proposed guidelines. Fed Reg 45:74984-74988.

Environmental Protection Agency (1982). Assessment of risks to human reproduction and to development of the human conceptus from exposure to environmental substances. Oak Ridge National Laboratory, ORNL/EIS-197, EPA-600/9-82-001.

Faustman-Watts E, Greenaway J, Namking M, Fantel A, Juchau M (1982). Teratogenicity in vitro of 2-acetylaminofluorene in the rat: role of biotransformation. Teratology in press.

Friedman RD (1979). Sensitive populations and environmental standards: a legal analysis. The Conservation Foundation, Washington DC.

Office of Technology Assessment (U.S. Congress) (1981). Assessment of technologies for determining cancer risks from the environment.

Omenn GS (1979). Genetics and epidemiology: medical interventions and public policy. Sociol Biology 26:117-125.

Omenn GS (1982). Predictive identification of hypersusceptible individuals. J Occ Med 24:369-374.

Omenn GS, Motulsky AG (1978). Eco-genetics: genetic variation in susceptibility to environmental agents. In Cohen BH, Lilienfeld AM, Huang PC (eds): "Genetic Issues in Public Health and Medicine," Springfield, Ill: C. C. Thomas, pp. 83-111.

Schull WJ, Otake M, Neel JV (1981). A reappraisal of the genetic effects of the atomic bombs: summary of a thirty-four year study. Science 213:1220-1227.

THE ROLE OF SURVEILLANCE IN MONITORING REPRODUCTIVE HEALTH

Michael J. Rosenberg, M.D., M.P.H.
William E. Halperin, M.D., M.P.H.

National Institute for Occupational
Safety and Health
Centers for Disease Control
Cincinnati, Ohio 45226

Introduction

John Evans of the World Bank points out that health systems have evolved in three phases over the past two centuries: the first phase dealt with infectious diseases liked to poverty, malnutrition, and poor personal hygiene; the second phase, with chronic diseases; and the third or current phase, is evolving in response to health hazards arising from environmental exposure to an increasing number of chemicals, drugs, and other toxic substances (1). Over the past decade in the United States, concerns for reproductive health are part of the third phase. Evidence for scientific interest in the field is the increasing number of publications each year which consider the effect of various exposures on reproduction (2).

This interest in reproductive health has been accompanied by several related occurrences in the United States. First, the average number of children in each family has been falling, and childbearing is being delayed longer than in the past (3), so the health of each child is becoming increasingly important. Second, the proportion of women in the workforce has increased steadily from 37% in 1970 to 42% in 1978 and is expected to reach 45% by 1990 (4), so more women will be exposed to occupational reproductive hazards. Today, the workforce of 109 million persons includes 77 million men and women between the ages of 16 and 44 years (4).

Today we face the question of how to reliably detect

environmental (including the working environment) exposures which lead to adverse reproductive outcomes. By the term adverse reproductive outcomes, we refer to the three broad categories of impaired fertility, fetal death, and birth or developmental defects.

Let us examine the three methods for detecting these hazards. The first is what we call observational epidemiology. This includes observations by worker or other groups and observations by astute clinicians. The second is clues which have been uncovered by toxicologists, working mainly with animals. The last is surveillance. For this talk, we will adopt the definition of surveillance proposed by Langmuir: "the continued watchfulness over the distribution and trends of incidence through the systematic collection, consolidation, and evaluation of morbidity and mortality reports and other relevant data" (5).

Defining reproductive hazards

Of these three approaches, observational epidemiology has clearly been the most successful at defining the hazards which are recognized today. Probably the best known example of a reproductive hazard being recognized through such efforts was in 1941, when Gregg noted a high prevalence of cataracts in children born of mothers with rubella (6). Other alert clinicians were first to raise what was subsequently confirmed as an association between thalidomide and limb-reduction defects (7), fetal alcohol syndrome (8), and diethylstilbestrol and clear cell adenocarcinoma (9). A striking illustration of workers first bringing a problem to light was provided when, in 1977, five workers from a plant manufacturing dibromochloropropane, or DBCP, reported voluntarily for evaluation of infertility. Their markedly reduced sperm counts and similar reductions in other exposed workers were subsequently demonstrated to be associated with duration and extent of exposure to DBCP (10-12).

Like any other system, the observational approach has also led researchers up blind alleys. In August of 1980, for example, a corporate physician reported an apparently increased rate of spontaneous abortions among female workers who used video display terminal (VDT's).

An investigation conducted by the Centers for Disease Control confirmed an excess of spontaneous abortions which were clustered in time but failed to find an association with VDT or other risk factors (13).

These examples raise several questions regarding the observational approach (and I include in our definition of the observational approach the resources necessary to further evaluate the initial questions with larger, more formal studies). What has the balance been between its successes (hazards identified) and the resources consumed in pursuit of what turned out to be blind alleys? Can we continue to rely on this serendipitous method to detect hazards, or do we need to add more systematic evaluation? Answering the first question is not presently possible, and the answer to the second depends on the perceived need to define such hazards.

The second broad means to identify reproductive hazards is the work of toxicologists. As we see in preceeding conference papers, the advantages to laboratory work include controlled exposures and evaluations generally which require far less time than do epidemiologic studies in man. Thus far, toxicology has played a limited role in determining which field studies of humans are conducted It is not the base that toxicologists have failed to identify reproductive toxins, but rather that these clues have not been pursued by study of exposed human populations. Toxicology's success stories are also stories of failure. DBCP is one such example. Toxicologists have linked DBCP with reproductive toxicity, but this finding remained in the toxicology literature until after its rediscovery when workers' complaints led to epidemiologic study. A true success story comes from waste anesthetic gases, which were targeted for further study by toxicologists and later confirmed as reproductive toxins in man. Toxicologists at the National Institute for Occupational Safety and Health (NIOSH) have discovered potent reproductive toxins which we are now searching for among groups of workers for field study evaluation. One such toxin with reproductive effects is 2-ethoxyethanol, a member of the family of glycol ethers, which in rats produces marked reproductive impairment with moderate exposures.

A systematic review of the animal literature by toxicologists as well as researchers from other disciplines which bear on human reproduction is needed to establish a priority ranking of agent most suitable for human studies. In addition to animal work, this effort should consider the extent of exposure for each agent and the feasibility of corrective actions if toxicity is identified.

Surveillance is the third method that could be used to identify reproductive hazards. There are two examples of systems which conduct reproductive surveillance in the United States today, the Birth Defects Monitoring Program (BDMP) and the Metropolitan Atlanta Congenital Defects Program (MACDP) (14). Both came into existence several years after the thalidomide tragedy, with the implicit mandate of recognizing and thus controlling such events more quickly.

These systems have shown that it is possible to measure the incidence of birth defects in a timely fashion and determine trends. As reviewed in a preceding conference paper (15), we know the time trends for some 150 categories of birth defects. These data have called attention to substantial increases of congenital heart disease, the excess risk for neural tube defects in the Appalachian Region, and the decrease in neural tube defects for the United States as a whole. There is little doubt that these surveillance systems can identify in a timely fashion changes in incidence of major birth defects.

These two data sets have provided the epidemiologic base upon which to build studies seeking new etiologic agents or to evaluate putative etiologic agents of public health significance. Dr. Oakley points out that case-control surveillance produced an association between diazepam and facial clefts (15). Dr. Legator called attention to associations with parental occupation which emerged from the data set (16).

There are several examples of how the data can be used to test hypothesis raised by others. When the association of vinyl chloride and neural tube defects was raised, the BDMP already had surveillance data on birth defects from hospitals in 2 countries with vinyl chloride plants. This data provided the CDC with the means to interview parents as part of a case-control study to rapidly test the hypothesis. When women were interviewed and the data analyzed, the authors concluded that the increase in neural tube defects in these communities was not due to occupational or environmental exposure to vinyl chloride (17).

Los Angeles investigators raised the issues that airport noise might cause neural tube defects (18). The birth defects surveillance data in Atlanta provided the base to test their hypothesis, which could not be confirmed (19).

A great deal of the cost of studies derives from identifying the populations to be studied. Since study samples are identified in the surveillance process, the analytical epidemiologic study could be performed on this data base at considerably reduced costs when compared with costs of newly initiated studies.

The two CDC birth defects surveililance programs have not been responsible for identifying a new "thalidomide", but fortunately, neither has any other system. The data have provide valuable information about long term trends, have raised etiologic hypothesis and have tested hypothesis raised by others of environmental exposures and birth defects.

In a reproductive setting, surveillance has other productive advantages. First, background rates or reproductive events such as spontaneous abortions permit defining risk in subpopulations. Approximately 15% of pregnancies end in spontaneous abortion (20,21), but this undoubtedly represents a mixture of higher and lower rates, just as do overall rates of cancer. Here we can draw an example from experience with cancer surveillance. After plotting the number of cases of lung cancer by county, several high-incidence areas stood out. Investigation of the high rates in coastal Georgia led to the identification of shipyard use of asbestos as a causative agent (22).

The availability of rates of spontaneous abortion and other reproductive events may also help with evaluation of clusters by defining whether there is a need for more extensive study.

Second, surveillance can monitor trends which may suggest research avenues and even provide clues. The BDMP has recently reflected a trend of increasing ventriculoseptal defects and patent ductus arteriosus and a decreasing trend of neural tube defects. The possible causes for these are unknown but are being investigated.

Routine monitoring of reproduction might include usual as well as unusual events. Information from such pregnancies could be used as historical controls for other studies, and a large pool of such information increases the potential statistical yield by close matching. Since reproductive studies generally require large populations, such use of historical controls may decrease the number of subjects needed and in some cases, as when too few exposed or case subjects are available, make such studies possible.

Surveillance can also generate hypothesis and help establish priorities for research. This can come from trends, as mentioned earlier, or from investigating associations at a single time. An example is provided by the BDMP, which raised the possible association between diazepam use and cleft lip. NIOSH has begun coding parental occupation on infants' death certificates in order to investigate the relationship between occupation and infant death. The ability to examine such associations is an important adjunct to animal studies and observations from clinicians and workers.

Fifth, such a system may identify additional associations of particular events. Such an example comes from vinyl chloride. When its association with angiosarcoma was noted, a registry for the tumor was establised. The registry confirmed the association of angiosarcoma with Thoratrast, a radioisotope solution used for liver-spleen scans, and with Fowler's solution, an arsenic-containing solution formerly used as a tonic. Both these association had been generally accepted on the basis of case reports, again by alert clinicians. However, association with a third drug, androgenic anabolic steroids, was discovered (23).

Finally, a system which monitors early fetal loss might be a more sensitive indicator of reproductive risk than later events such as birth defects. If reproductive risk is considered a continuum extending from impaired fertility to fetal death to birth defects to developmental defects, stronger toxins may, for example, increase the number of spontaneous abortions and thus decrease the number of children with birth defects available for study.

Problems with surveillance

The first question is what to study. What we can study most centrally depends on completeness of ascertainment. Thus studying events such as birth, for which medical attention is essentially universal, is relatively easy; the success of the BDMP and programs which determine infant death is evidence. Since a woman's likelihood of seeking medical attention for problems of pregnancy increases as the length of gestation, monitoring of spontaneous abortions might be the first goal of a surveillance system. Such a system might first seek to monitor all pregnancies which extend beyond 28 weeks, since this cutoff is presently used by New York and several other states and results in greater than 90% ascertainment (24). Since fewer than 10% of spontaneous abortions occur after this time however, the yield from such a system would be limited to small numbers. Similarly, reliable monitoring of infertility is difficult because of unpredictable likelihood of seeking medical evaluation and the lack of standardized reporting.

A surveillance system must incorporate periodic evaluation. We must assure that the information collected is useful, as judged by its ultimately inducing change, and that the information is not otherwise available. Minimal criteria for completeness and acuracy of data collection, timeliness, and accuracy of processing and dissemination should be established and compared with the system's operations.

Conclusions

Reproductive studies in man are difficult because of large samples needed and the real possibility that effects are subtle compared to common exposures such as alcohol and smoking. This means that studies in man must be carefully considered, for resources to systematically explore the relationship between environment and reproductive hazards. Of the three methods available (observational, toxicology, and surveillance), the most productive in identifying hazards has been the observational. However, we are not currently able to weigh these benefits with the costs as measured by resources used in evaluating possible hazards. For demonstrating a lack of association, surveillance coupled with case-control studies has been most effective. Animal work has been too recent to evaluate. For reproductive monitoring, a carefully designed and instituted surveillance effort stands to provide the ability to systematically explore the relation between the environment and reproduction. Surveillance will be a central component in that exploration, because, in Langmuir's words, "Good surveillance does not necessarily ensure making of right decisions, but it reduces the chances of wrong ones" (6).

REFERENCES

1. Evans JR, Hall KL, Warford J: Health care in the developing world: Problems of scarcity and choice. New Engl J Med 1981; 305:1117-28.

2. Strobino BL, Kline J, Stein Z: Summary of published data and an annotated bibliography on exposures and reproductive function. In Bloom AD, editor: Guidelines for studies of human populations exposed to mutagenic and reproductive hazards. March of Dimes Birth Defect Foundation, New York, 1981.

3. Fertility of American Women: June 1981. Current Population Reprints Series P-20, No. 369, Bureau of the Census, 1982.

4. A statistical portrait of women in the United States: 1978. Current Population Reprints Special Studies Series 1-23, No. 100, Bureau of the Census, 1980.

5. Langmuir AD: The surveillance of communicable diseases of national importance. New Engl J Med 1963; 268:182-02.

6. Gregg NM. Congenital cataracts following German measles in the mother. Tr Ophth Soc of Austr 1941; 3:35-39.

7. McBride WG. Thalidomide and congenital abnormalities. Lancet 1961; 2:1358.

8. Lemoine P, Harousseas H., Boreyru JP, et al. Les infants de parents alcoholiques: Anomalies observees, a

propos de 127 cas. Quest Med 1968; 25: 476-82.

9. Herbst AL, Ulfelder H., Poskanzer D.C. Adenocarcinoma of the vagina: Association of maternal stilbestrol therapy with tumor appearance in young women. N Eng J Med 1971; 284:878-81.

10. Whorton DM, Krauss RM, Marshall S, et al. Infertility in male pesticide workers. Lancet 1977; 2:1259-61.

11. Whorton D, Milby TH, Krauss R, et al. Testicular function in DBCP exposed pesiticide workers. J Occup Med 1979: 21:161-6.

12. Glass RI, Lyness RN, Mengle DC, et al. Sperm count depression in pesticide applicators exposed to dibromochloropropane. Am J Epid 1979, 109.346-57.

13. Binkin NJ. Personal communication.

14. Edmonds LD, Layde PM, James LM, et al. Congenital malformation surveillance: Two American Systems. Int J Epid 1981; 10:247-52.

15. Safra MJ, Oakley GP. Assocation between cleft lip with or without cleft palate and prenatal exposure to diazepam. Lancet 1975; 2:478-80.

16. Legator MP. Conference Review to be filled in.

17. Edmonds LD, Anderson LE, Flynt JW, et al. Congenital central nervous sytem malformations in vinyl chloride monomer exposure: A community study. Teratology 1978; 17:137-42.

18. Jones FL, Tauscher J. Residence under an airport landing pattern as a factor in teratism. Arch Environ Health 1978; 33:10-12.

19. Edmonds LD, Layde PM, Erichson JD. Airport noise and tertogenesis. Arch Environ Health 1979; 34:243-7.

20. French FE, Bierman JM: Probabilities of fetal mortality Publ Hlth Rep 1962; 77:835-45.

21. Harlap S, Shiono PH, Ramcharan S. A life table of

spontaneous abortions and the effects of age, parity and other variables. In Hood EG, Porter I, eds, Reproductive loss. 1980; Academic Press New York.

22. Blot WJ, Harrington MJ, Toledo A, et al. Lung cancer after employment in shipyards during World War II. New Engl J Med 1978; 299:620-4.

23. Falk H, Thomas LB, Popper H, et al. Heptic angio-sarcoma asociated with androgenic-anabolic steroids. Lancet 1979; 2:1120-3.

24. Quality control vital records processing. Vital Statisitics Review, New York, State of New York Department of Health, 1976.

FIELD STUDIES: LESSONS LEARNED

Donald Whorton, M.D.

Occupational studies of male reproductive effects were essentially non-existent in the U.S.A. until mid-1977 with the discovery of sterility and infertility among pesticide formulation workers of a plant in Central California. Within a short period of time, the causative agent was found to be dibromochloroproprane (DBCP). During 1977 and 1978, six other studies on the effect os DBCP were done. All had similar results to the initial California study.

Since that initial discovery, numerous studies of workplace exposures to other agents have been undertaken. Most studies have been negative or contradictory. None have been as dramatically positive as DBCP. The results from each of these other studies (as known to the author) will be presented.

Problems with such studies (beyond the usual occupational problems of determining extent of exposure) include participation rate, difficulty obtaining concurrent controls, and representatives of comparison populations.

Sperm count has been the most widely used semen characteristic in these studies. In situations when concurrent controls cannot be obtained, historical controls have been used. Problems in interpretation of some studies, e.g., ethylene dibromide, have occurred in choosing the representativeness of historical controls. In sperm count results, lack of control for continence time can cause wide variation in results. Data showing these variations will be presented.

While semen analyses can evaluate the external appearance and number of sperm cells, determination of the potential mutagenic effects on sperm cells is more difficult. While one can measure double Y bodies, the precise meaning of this test is unclear. Another method is to evaluate the rate of spontaneous abortions among the wives of the male employees. An accurate study of this parameter would include obtaining reproductive histories from the women, as well as accurate work history (time of exposure) from the men. Data will be presented which demonstrate the differences in reporting of the men compared to their wives for spontaneous abortion (number and timing) and number of

© 1984 Alan R. Liss, Inc.

children. Only 77% of reported spontaneous abortions were actually spontaneous abortions when medical records were reviewed for vertification.

I want to thank the organizers for inviting me I am going to discuss my impressions of the current status of occupational reproductive data, and some of the problems in doing field studies. I will emphasize studies of the male worker as I have more experience with them. I will discuss problems with sperm counts and two studies of spontaneous abortions in wives of exposed male workers.

It is easy to say how many occupational studies of testicular function has been done in the United States prior to 1977. There were none. Thus, one would classify all occupational reproductive problems as a new area of scientific endeavor. The issue of reproductive problems first arose at the federal level in the OSHA lead standard. There were several reasons for that. Historically, the industry had prohibited women from work in lead areas. In addition, there was a study showing adverse effects on semen quality of male workers in Romania.

The inclusion of testicular function data in the standard making process raised many questions. Could such studies be done in the United states? Could one go to a workplace, hand out specimen containers and request the men return them with semen samples? Would such a request be dangerous to the health of the investigator?

Since collecting semen is a noninvasive technique, it should theoretically be easy to obtain. But in practice, one must contend with the male ego and social mores. Obtaining blood samples is orders of magnitude easier than collecting semen samples.

In 1977, in July, I received semen results of seven men employed at a chemical company in California. The issue of infertility was perceived by the men themselves. The semen results were sent to me as I happened to have a relationship with the union. This was the beginning of the human investigations into DBCP which proved that the adverse reproductive effects could also affect males.

Table I lists the toxic effects of DBCP (1). A part of the report by Torkelson et al was a subchronic feeding study that showed testicular atrophy in rats, rabbits and guinea pigs. More importantly, it showed testicular atrophy at levels in which no other effects were seen. The lesson from this was: If in a subacute or chronic study, the only adverse effect observed is testicular, one should pay attention to it. In the 1970s, data showed DBCP was an animal carcinogen and an in vitro mutagen. Since 1977, there have been a variety of studies done on infertility and sterility in men.

The first human study was published by us in 1977 Lancet (2). We were fortunate that in the first study done in the U.S., we had 100% participation. No other study requiring semen samples has had such a participation rate. It is difficult to get an entire population at risk to participate. Of the 36 men examined, 11 had vasectomies, which was our first surprise. We had no idea the vasectomy rate in California was so high. (In another study, we found 40% of the men were vasectomized.) The 25 non-vasectomized men fell into 3 groups: 11 long-term exposed (more than 3 years), 1 in the minimal-exposed (less than 3 months) and 3 in-between (one year). For the long-term exposed the mean sperm count was 200,000 per ml. Nine of these eleven were azoospermic (zero sperm count). For the short-term exposed group of 11, the mean sperm count was 92 per ml. The Follicle Stimulating Hormone (FSH) and Luteinizing Hormone (LH) were statistically significantly elevated for the long-term exposed compared to the short-term exposed.

Table II is a summary of all the studies done in 1977-1978 and reported in English (3). Some of the studies have been published in peer-reviewed journals while others have only been reported at conferences. Oxy Chem was a pesticide formulator. Shell in Colorado stopped producing DBCP in January 1976. Shell in Alabama had been producing DBCP since January 1976. Dow in Arkansas has been producing DBCP for about 2 1/2 years. THe CDC study was done on pesticide applicators in California. The Israeli study was done on producers. The PEA study was done on pesticide applicators and salesmen in the South by the University of North Carolina.

Table I

DBCP (1,2-dibromochloro-3-propane)

Use: -Nematocide

Common Crops:
  U.S.:   -Citrus, Grapes, Peaches, Pineapple, Tomatoes, Soybeans, Etc.
Central
America
and Israel: -Bananas

Properties: -Liquid

Toxicities: -Mild irritant, hepatic and renal toxin, testicular atrophy in laboratory animals (1961).
-Animal carcinogen (1973).
-In Vitro mutagen (1975).
-Infertility and sterility in man (1977).

Table II

Summary of DBCP Studies 1977-1978

Study	No. of Subjects Examined *	Number of Men with Sperm Count Results in Millions per ml		
		0	.1-9	10-19
Oxy Chem (California)	114	15	8	12
Shell (Colorado)	64	5	2	7
Shell (Alabama)	71	1	6	5
Dow (Arkansas)	86	30	(17)**	(3)
California Applicators (CDC)	74	6	8	(7)
Israel	23	12	6	0
EPA	53	2	12	8
TOTALS	485	71	59	42
%		14.6	12.2	8.7

* Variable exposures
** Numbers in parentheses extrapolated from data presented.

Of the 485 men examined, $14\frac{1}{2}$ % were azoospermic and another 21% were oligospermic (sperm count less than 20 million per ml). Thus, more than 1/3 of the studied population had low or zero sperm counts. One should remember that each study had different populations, methods, and participation rates.

No other chemical agent has been found to produce the dramatic adverse effects of DBCP on testicular function. Some agents have shown marginal and thus debatable effects: e.g., ethylene dibromide, Kepone, lead, and toluene diamine.

Ethylene dilbromide is discussed later in this presentation. There is a report that the individuals who were severely poisoned with Kepone also had low sperm counts. The lead study has been discussed. The data on toluene diamine are lacking in enough numbers to show statistical significance, but there is an apparent reduction in sperm count in the heavily exposed group.

Other agents with variable results are anesthetic gases, carbon disulfide and cabaryl. Epidemiologic studies have shown an increase in birth defects among offspring of male anesthesiologists. Semen data for anesthesiologists are normal.

A study on carbon disulfide showed abnormal semen quality in Romania in the early 1970s. A similar study done in the U.S. by Dr. Channing Meyer showed no abnormalities. The differences are probably due to differences in exposure. Carbaryl was discussed yesterday by Dr. Wyrobek.

Chemical agents or compounds with negative results include epichlorohydrin, polybrominated biphenyls, paratertiary butyl benzoic acid, and glycerin products (4).

Studies currently being conducted or just completed but not yet published include two studies on pineapple workers exposed to DBCP. One, a cross-sectional study done as a NIOSH HHE, showed no difference between exposed and non-exposed. The second study, also a NIOSH HHE, is a longitudinal study conducted over the planting season. No difference in sperm count or morphology were observed between exposed and controls. Dr. Wyrobek has done a study on a few lead workers in Pennsylvania. No effect was observed. In addition, a study done in Sweden with lead workers having relatively low blood levels showed no effect. A study done for American Hospital Supply Company on ethylene oxide exposed workers reported no adverse effects, although the oligospermic rate was about 20%. It is my understanding that Dow corporation conducted a study on their employees exposed to glycol ethers. The study has been reported as negative. I have been involved in evaluated men exposed to steroid compounds at a chemical plant in the Bahamas. No adverse effects were found.

What are the problems in conducting studies on testicular function among male employees? Obviously, participation rate is a big issue. How does one obtain participation? It is not easy. Anybody who thinks it is easy has not done such a study. While in our first DBCP study, we had 100% participation (36 men), that is a highly unusual situation. In no other study has anyone obtained such a participation rate. The second issue involves who one is to use for controls. How or can one obtain concurrent controls in the plant? If there is reluctance in exposed men to participate, there is greater reluctance by the non-exposed to participate. If one cannot obtain an adequate number of concurrent controls, selection of an appropriate comparison population is vital. If one uses an inappropriate comparison population, positive data for depressed sperm counts become negative.

Data analysis methods are somewhat more difficult, a sperm counts and possibly sperm morphology are not normally distributed. Even if one attempts to normalize by using the square root of log of the sperm count, this still does not always give a normal distribution. The use of non-parametric statistics becomes necessary; thus one needs access to a statistician if s/he does not already possess such skills.

The power of a negative study was covered in yesterday's discussion. This issue is important in evaluating semen parameters. Is the study negative statistically only because the number of subjects was inadequate? As was discussed yesterday, morphology appears to be a more sensitive measure than count, based on numbers required to see small differences. That may be true for some compounds, but may not be true for others. The DBCP study is an example in which sperm count appears to be the more sensitive parameter. In the mouse, the morphology remains unaffected while the count decreases. Thus, if the agent is germ cell toxic, perhaps morphology is not more sensitive than count. If the agent affects some other phase of spermatogenesis, them morphology may be more sensitive. I do not think we have enough information to answer this question yet.

How does one evaluate potential mutagenic effects in sperm cells? We heard yesterday about the Y body test. The double Y is assumed to be two Y chromosomes, but this is not proven. The other method would be to evaluate the pregnancy outcomes of spontaneous abortions and birth defects. I will discuss this later.

The evaluation of sperm motility poses many problems. Motility determination is time and temperature dependent. The semen needs to be kept at least at room temperature (21 C) if not warmer, should not be placed in a refrigerator, and needs to be examined within two hours of ejaculation. There are logistical difficulties in doing motility examinations in occupational settings compared to clinical or physician's office settings. In most occupatonal studies, the best method for obtaining semen samples is to have them collected at home and brought fresh to a workplace collection point. One needs the laboratory technicians to be set up to do the analysis at the workplace. If the men work multiple shifts, this arrangement can be very expensive. There are drawbacks to having the semen samples collected at work. The room used for such a purpose can cause considerable disruption and inhibit participation. I know of one manager where collection of semen samples at work was proposed (not by me) whose reaction was he was not going to have any (expletives deleted) ejaculation rooms in his plant.

Should one exclude men with certain medical histories or physical findings from the study (e.g., men with varicoceles)? Perhaps they are at higher risk for the toxic effect than men who do not have this condition. Obviously, one must exclude vasectomized men. One must be very careful about not excluding the high risk population from the data analysis. If one is considering blood test, FSH, LHN, or testosterone as screening tests, they have been found to be too insensitive unless the sperm counts are zero or close to it.

Studies requiring semen samples necessitate full cooperation and participation by the employee. Collecting the sample is not a passive activity. The employees must be sufficiently motivated about the worthiness of the study. There are many reasons for individuals to decline to participate. Obviously, if they had a previous vasectomy, they do not need to participate. Many men whose wives are unable to have children (menopause, surgical sterilization, etc.) do not feel this type of study is useful for them. In addition, in my experience, frequently men who are beyond age 50 say, "Hey, Doc, I have a difficult enough time getting an erection as it is, and I don't want to waste it on this."

There is a real question in interpreting semen results with respect to fertility. What is a low count? How does one explain a sperm count of 10, 15 or 20 million per ml? One must be ready to discuss the meaning of results as one will find men with low counts or other abnormalities in any population. Follow-up evaluations need to be arranged prior to conducting the study. Some men will be very insistent about follow-up. The man often will not only want you to follow him up, but will want immediate treatment to make him normal tomorrow.

What is the rate of oligospermia among men in the U.S.? Oligospermia is currently defined as less than 20 million spermatozoa per ml of ejacuate. Table III shows data from four published studies, plus data we have collected. All the data have been published or collected during the past 8 years. Three of the studies evaluated sperm count from men at vasectomy clinics. The results of oligospermia are strikingly different. In two of the studies, 20 and 23% of the men were oligospermic (5,8), yet in the third clinic only 7% were oligospermic (6). The one difference among these populations was continence time. The study by Rehan required 48-hour minimum, while the other two studies did not. One must remember that men obtaining a vasectomy are told that more ejaculations prior to the vasectomy, the quicker the sperm count will become zero (the intended purpose of the procedure). Another factor in ejaculatory acivity prior to the operation is the perceived soreness afterwards, which will put some men out of commission for a while. Thus, several factors play a role in prompting numerous ejaculations prior the the operation.

The data from the prenatal clinic (7) are similar to the one vasectomy clinic data. The final data set in the table are the composite of men from seven of our studies. The men were either unexposed controls or were exposed subjects in which no adverse testicular effects were observed. Criteria for inclusion were no history of a vasectomy (including vasovasotomy) and at least 48-hour continence time. Only 7.5% of the population were oligospermic. All of these men were taken from worker populations.

Table III

Data from Last Eight Years on Apparently Fertile Men

Sperm Count Millions/ml	Nelson & Bunge ('74)	Rehan et al ('75)	Sobrero & Rehan ('75)	Smith et al ('81)	EHA ('82)
<20	20	7	5	23	7.5
20-40	31	16	13	22	15.0
40-100	42	52	62	34	46.0
>100	7	15	20	21	39.0
N	386	1300	100	4435	335.0
Median	38	65	68	45	78.0
Mean	48	79	81	66	95.0
Source	Pre-vasectomy clinic	Pre-vasectomy clinic	Pre-natal clinic	Pre-vasectomy clinic	Workers

Impotence can be another major problem in older men, as can medical disorders like a retrograde ejaculation. There are individuals who refuse to participate because they are afraid of an abnormal result. One 20-year-old refused to participate, as he was afraid if he even knew he had a low sperm count, no woman would ever want to marry him; thus he felt it was best not to know. There are religious and/or cultural reasons among some groups. The Bible prohibits masturbtion for whatever reason. Religious reasons have not been a major difficulty in the U.S. Lastly, some men find the type of study too personal. They do not want anyone to know about their semen quality.

There have been attempts to increase participation by offering some type of financial reward to the employee - either cash, increased vacation time, etc. There are arguments for and against such incentives.

An important variable to control is time since last ejaculation (continence time). This time is usually set, somewhat arbitrarily at 48-72 or 96 hours. Ejaculation frequency can also be important to know in some cases. For example, one man in early DBCP investigations had a sperm count of 18 million per ml. A repeat sample also gave a similar result. He was quite upset. I saw him 5 months later and his sperm count was 100 million per ml. Discussion with him revealed that at the time of the initial two counts, he was recently married and having sexual intercourse 5 times per day. He waited the 48 hours abstinence time to the minute. Five months later, he was having sex about once every three days and he waited five days.

Masturbation is the preferred method of collecting the sample. However, coitus interruptus can be used providing the entire ejaculation is caught. In most cases, the intitial portion of the ejaculate contains most of the sperm cells. In another case, a man used coitus interruptus to collect the sample. He wrote on the container label he did not think it was a good sample. The count was low. Upon repeat sampling, he had a repeat "bad sample" and a low count. I learned after talking to him that coitus interruptus was done the first time with his wife and the second time with his girl-friend. I finally convinced him to masturbate so a proper sample could be evaluated. The results were normospermic the third time.

Another major difference among these populations is median and mean sperm count. As an example of the problem in not controlling for continence time, the median sperm count of 45 million per ml as reported by Smith is the same as our ever exposed DBCP population in California, as reported in 1979.

All of these data show problems inherent in using external data to compare results of a study without an adequate control group.

The use of external comparison groups has practical importance in California. A small insect, the Mediterranean fruit fly, caused havoc in California for several years. Fruit from certain areas of the state had to be fumigated. A popular fumigant was ethylene dibromide (EDB).

EDB is closely related to dibromochloropropane (DBCP). Both can cause similar toxic effects in animal and *in vitro* testing. There were two studies done in 1977 on workers exposed to EDB: one in Louisiana and one in Texas. Data from both were presented in October 1977 at a NIOSH conference on DBCP.

The Lousiana study had 59 exposed participants without any controls. The California Department of Health obtained the raw sperm count data for all 59 men. The median sperm count was 57 million per ml. The men in this study had been divided into a low-exposed group (<0.5 ppm) and a higher exposed group (>0.5-5.0 ppm) Unfortunately, the key to which men belonged in which group was lost. The data presented at the NIOSH conference in 1977, however, indicated there was a difference between these two subgroups. A rough estimate of the median sperm counts would be 45-50 million per ml for the higher exposed and 65 million per ml for the low-exposed group.

In the Texas study, 17 men in exposed jobs without a previous history of sterility or infertility provided semen samples; however, only 12 were included in the analyses. The other five were omitted due to continence time of less than 24 hours or of unknown duration. The median sperm count was 54 million per ml. Six controls from the same plant with the same criteria had a median count of 79 million per ml.

In my opinion, these two studies would indicate some reduction in sperm count had occurred in the men in the exposed job categories. Unfortunately, without the ability to separate men into the exposure categories in the Louisiana study, further analysis is difficult. However, these studies do show that the utilization of a vasectomy clinic population without any continence time requirement can be very misleading.

Another problem with sperm counts can be the laboratory. While sperm counts are a procedure that can be done in most laboratories, the question of accuracy is important. Table IV figures are from a study I have been involved in, the results of six men tested by Laboratory A and later by Laboratory B. In general, there is close to a 10-fold difference (except the azoospermic man). Laboratory A had problems in their dilution calculations. This only points out the need for caution in accepting laboratory results.

Another commonly evaluated semen characteristic is spermatozoa morphology. Difficulties with this evaluation is comparability of results among laboratories. There is general good intra-laboratory agreement, but poor inter-laboratory agreement. The issue is definition of normal and categories of abnormalities. There are no accepted population data in the U.S. for normal morphology. One method to evaluate mutagenic effects on spermatogenesis is to study spontaneous abortions among the wives of male employees. We have been involved in such a study. We obtained separate histories form the men and their wives about the number of joint pregnancies. We found differences in number and timing of spontaneous abortions as reported by the husband and wife. In addition, we found differences in number of children. The men reported 157 children, while their wives reported 165. The women reported 27 spontaneous abortions; the men reported 24. When the spontaneous abortions reported by the wives and their husbands were compared, there was agreement on only 21. Of the 21 agreed-upon spontaneous abortions, there was agreement on time of event (within one month) in only three cases. The others varied up to 20 months. Even among the women, the more distant the event, the poorer the recall concerning time.

Table IV

Comparison of Sperm Counts on Six Men
Done in Two Different Laboratories

Sperm Counts (Millions/ml)

Male tested	Laboratory A	Laboratory B
#1	0	0
#2	<1	3
#3	4	63
#4	17	158
#5	18	82
#6	27	238

EHA Data, 1982

We sought verification of the reported spontaneous abortions by examining the women's medical records. There were 26 reported spontaneous abortions for which we obtained medical records. We accepted one or more of the following as verification of pregnancy: products of conception at the time of spontaneous abortion, a positive pregnancy test reported in the medical records, and/or declaration of pregnancy based on an examination by a physisian. Of the 26 reported spontaneous abortions, only 20 were actual spontaneous abortions. The other 6 had some form of dysfunctional bleeding and were not or had not been pregnant.

A study of spontneous abortions of wives of DBCP exposed banana farmers in Israel examined medical records of the women and compared the number of pregnancies, number of live births, and number of spontaneous abortions before and after husband exposure to DBCP. A three-fold increase in spontneous abortions occurred after exposure to DBCP (10). This is the only study on DBCP exposure in humans that measures effects other than sperm count.

Within the scientific process is critical review by peers. However, this review can sometimes miss the target. The following appeared in the NIOSH Current Intelligence Bulletin 37 (1981):

> In studies of the effect of EDB exposure on sperm production conducted in 1977 and 1978, sperm counts were performed on 59 workers potentially exposed to EDB in a chemical manufacturing plant. The author concluded that exposures EDB at the levels found for this population (less than 5.0 ppm) had no adverse effect on sperm counts. Neither possible confounding exposures nor the health of the individuals was discussed. No tests were performed to determine sperm motility, penetrance, morphology, or density or to evaluate other sperm function parameters. [emphasis added]

I direct your attention to the first sentence and last sentence. In the first sentence, there is an acknowledgement that sperm counts were performed on 59 workers. However, in the last sentence, there is criticism that no tests were performed to test sperm motility, penetrance, morphology, density, or to evaluate other sperm function parameters. Sperm density (third sentence) and sperm count (first sentence) are synonymous terms. In addition, as discussed yesterday, penetrance studies are not readily available or interpretable clinical tests. Field studies are difficult enough to do without being criticized inappropriately and incorrectly. Reviewers should remember field studies are not metabolic wands as any one with experience in field studies can readily attest.

Questions and Answers

(1) I am concerned that the diagnosis in medical records may be less accurate than maternal recall of spontaneous abortions.

We adopted the position we needed objective evidence of pregnancy. As I said previously, that objective evidence could be a positive pregnancy test, physical examination, or products of conception at the time of the spontaneous abortion. Thus, we were not limiting ourselves to hospital or clinic cases with decidual tissue. One of the difficulties in any spontaneous abortion study one must address is definition. We could have missed early events which occurred, but the assumption is that this bias would not be different before or after exposure.

(2) Given all these difficulties in semen evaluation, do you see any place for semen evaluation in routine screening or routine surveillance of large employee populations?

I would like to say yes. I cannot. Except in special circumstances, you are not going to get continued participation by individuals. In addition, what will you do to individuals who will not or cannot participate? I think social factors will prohibit the routine use of semen in surveillance.

(3) In Israel, DCBP abortion studies a new way of any possible problem? (sic)

I am not sure I understand the question. I have not talked to the authors, but the data can raise some questions. The pre-exposure spontaneous abortion rate was only 6%, which is low. There is the obvious question of artifact. The other obvious issue is how well were they able to evaluate exposure times. While there may be criticisms of the study, it has the only data in town.

(4) We all know epidemiology is not as refined as lab studies, but nonetheless, with all the problems you can use it to find out real information. I think it can be used for the study of spontaneous abortions as well. I would like to just throw out to you our experience. We examined a workplace and found a significant cluster of spontaneous abortions in relation to risk from a particular chemical exposure. We obtained the information from men. Subsequently, we went back to their wives asking them to correct that information and then from the wives' responses, we went back to housefold records where about 75% of all spontaneous abortions that had occurred to individuals at this plant after working at the plant. Review of hospital records succeeded in cleaning of the data a little bit, but the overall pattern was virtually unchanged even with better information as to the timing of the event in relation to exposure. The cluster was still there, it was still significantly different than expected. To summarize, in a cluster of spontaneous abortions taken by men and verified by wives, look at medical records, even though the medical records did not change the total outcome.

In our study, including or excluding the six "non abortions" does not change the outcome. The primary reason for discussing the exclusion was to raise the issue of definition of spontaneous abortion which, as you know, is a major issue in spontaneous abortion studies. In addition, objective evidence may produce different results than subjective evidence. In the case of workplace exposures, knowledge of the timing of the event and the exposure history is important. Objective data are generally more reliable than recall, especially for remote events. Finally, in spontaneous abortion studies, one should not rely totally on the histories from the husbands.

(5) I am not clear on the rationale for recording sperm counts as cells/cc rather than total counts. It seemed to me, given the variation in volume, that it would make you get better sensitivity if counts were recorded as just total counts.

From what I understand of fertility, concentration may be more important than total count. For example, take two men with a total count of 100 millon sperm cells per ejaculate. If one man has a volume of one ml, then he has a count of 100 million per ml and would not expect to have a fertility problem based on count. On the other hand, if the second man had a volume of 10 ml, then he has 10 million sperm cells per ml. He would be considered oligospermic and may well have fertility difficulties. Finally, much of the data in the literature utilize the concentration measurements.

References

1. Whorton, MD: The Effects of Occupation on Male Reproductive Effects,, <u>Les Facteurs de la Fertilite Humaine</u> (1982), edited by A. Spira and P. Jouannet, Institut National de la Sante et de la Recherche Medicale, Paris, pp. 339-345.

2. Whorton, MD, Krauss, RM, Marshall, S and Milby, TH: Chemical Induced Infertility among Employees in a Pesticide Formulation Facility (1977). <u>Lancet</u> 2:1259-1261.

3. Whorton, MD: The effects of Occupation on Male Reproductive Effects, <u>Les Facteurs de la Fertilite Humaine</u> (1982), edited by A. Spira and P. Jouannet, Institut National de la Recherche Medicale, Paris, pp.339-345.

4. Whorton, MD: The effects of Occupation on Male Reproductive Effects, <u>Les Facteurs de la Fertilite Humaine</u> (1982) edited by A. Spira and P. Jouannet, Institut National de la Recherche Medicale, Paris, pp.339-345.

5. Nelson, CMK, and Bunge, RG: Fertil. Steril. (1974), 25:503-507.

6. Rehan, NE, Sobrero, AJ, and Fertig, JW: <u>Fertil Steril.</u> (1975), 26:492-502.

7. Sobrero, AJ, and Rehan, NE: The Semen of Fertile Men. II. Semen Characterisitics of 100 Fertile Men <u>Fertil. Steril.</u> (1975), 26:1048-1056.

8. Smith, KD: Statement to Department of Industrial Relations, Occupational Health Standards Board, November, 1981, p. 9

9. Whorton, MD, MIlby, TH, Krauss RM and HA Stubbs: Testicular Function in DBCP-Exposed Pesticide Workers, <u>J Occup Med</u> 21:161-166.

10. Kharrazi, M, Potashnik, G, and Goldsmith, JR: Reproductive Effects of Dibromochloropropane, <u>Israeli J Med Sci</u> (1980), 10: 403-406.

THE EFFECT OF COMMON EXPOSURES ON REPRODUCTIVE OUTCOMES

Carol J. R. Hogue, M.P.H., Ph.D.

Chief, Pregnancy Epidemiology Branch
Centers for Disease Control
Center for Health Promotion and Education
Division of Reproductive Health
Atlanta, GA   30333

The environment of the fertile couple and the pregnant woman includes several known risks to reproductive function--risks not limited to chemical exposure, but including psychological and physiological as well. Exposures to such multiple risks increase the complexity of epidemiologic investigation because of the need to evaluate environmental hazards within a milieu of exposure to other substances, of which several are known to affect reproductive outcome. This complexity can be handled methodologically and must be dealt with properly for adequate research to be conducted.

This chapter will present some of the ramifications for research on multiple risk factors. First, rates of adverse pregnancy outcomes associated with common hazards to reproductive function differ among groups of parents classified according to their level of exposure. Second, if one or more of these common reproductive hazards is a confounding factor in a study of an occupational or environmental chemical, investigators may encounter problems with data interpretation. Third, a common exposure may produce a synergistic effect with an environmental agent. Fourth, the investigator may have problems documenting common human exposures. Finally, sample size must be adequate to account for the effect of common exposures.

© 1984 Alan R. Liss, Inc.

ADVERSE EFFECTS OF COMMON EXPOSURES

It is well known that personal habits may affect reproductive outcome. Three very common exposures--smoking, alcohol consumption, and caffeine consumption--are listed in Table 1, along with associated adverse effects on reproduction. We have insufficient information to fill in all of the cells of this grid, but what we do know about the effects of maternal smoking and alcohol drinking is alarming. If a woman smokes or drinks during her pregnancy, she is exposing her fetus to increased risks of spontaneous abortion, low birth weight, perinatal mortality, and possibly, developmental defects. Drinking seven or more cups of coffee a day has been associated with increased risks of low birth weight (Hogue 1981). The effects of paternal exposure to smoking, alcohol, and coffee drinking have not been studied as extensively. At least one study of chronic alcohol drinking has indicated testicular atrophy, azoospermia, and testicular pathology (Turner, et al. 1977).

Table 1 Common Habitual Risks to Reproduction

	MATERNAL			PATERNAL		
	SM	AL	CA	SM	AL	CA
Infertility	?	?	?	?	X	?
Spontaneous Abortion	X	X	?	?	?	?
Low Birth Weight	X	X	X	0	0	?
Perinatal Mortality	X	X	?	?	?	?
Developmental Defects	X	X	?	?	?	?

SM = smoking; AL = alcohol; CA = caffeine.
X = effect found; 0 = effect not found;
? = effect not examined.

The remainder of this section focuses on known maternal risks. Paternal exposures may account for a couple's infertility or may produce effects on the germ cell resulting in chromosomal anomalies that cause spontaneous abortion or surviving infants with birth defects. Maternal exposures may produce similar effects; additionally, maternal exposures during pregnancy may cause teratogenic effects. Hence, one reason that mothers have been studied more is the broader range of effects and of time that maternal exposures have to produce effects. Another reason is that mothers have been more accessible for epidemiologic investigations. Now that reproductive epidemiology is moving into the workplace, paternal exposures should be the subject of more frequent study.

Other common exposures and risks to humans include drugs taken during pregnancy, dietary deficiencies or excesses, and certain physical characteristics and exposures. Fortunately, most drugs commonly taken during pregnancy, such as Bendectin™ and salicylates, have not been found to influence reproductive outcome in humans. On the other hand, the tragic effects of Thalidomide™ taught us not to be complacent about medications that pregnant women take. Furthermore, there could be pregnancy-related, metabolic interactions with drugs and environmental chemical exposures. Such interactions could be either protective or harmful.

With respect to dietary patterns, we have only begun to understand the impact diet may have on reproductive function. We know, for example, that maternal starvation increases the risk of adverse pregnancy outcomes (Stein, Susser 1975). Yet, clinical trials of protein supplementation for undernourished mothers have had little, if any, effect on improving pregnancy outcome (Rush, Stein, Susser 1980). Deficiencies of certain dietary components, such as zinc and folate, seem to increase the risk for neural tube defects (Flynn, et al. 1981; Laurence, et al. 1981). One curious feature of all maternal folate studies is that while significant differences in mean folate concentrations are found, folate levels for most mothers of infants with neural tube defects fall within normal ranges. There is seldom a clinically definable maternal folate deficiency when the child is born with a neural tube defect (Laurence, et al. 1981). Maternal folate deficiency is a known teratogen in rats (Nelson 1960), and the folate antagonist aminopterin has teratogenic effects in humans including anencephaly (Thiersch 1963). Hypervitaminoses, such as hypervitaminosis A, may also be teratogens in humans (Gal, et al. 1972).

Maternal weight gain is a controversial issue, although one factor which seems to be a risk regardless of maternal nutritional status is low prepregnancy maternal weight. As a matter of fact, low prepregnancy maternal weight may increase the risk of infertility as well, although the evidence for a fertility effect is not very strong (Bongaarts 1980).

Other maternal physical factors which affect reproductive function include parity (number of previous live births) and age. Nulliparous women are at higher risks of adverse pregnancy outcomes such as low birth weight and perinatal mortality than women who have had previous live births (Shapiro, et al. 1968). The parity-specific incidence of many pregnancy outcome variables follows a J-shaped curve, with highest risks in the first pregnancy, lowest risks in the second pregnancy, and a slow rise in adverse rates as parity increases after two. Very young nulliparous women are at particularly elevated risks of delivering low-weight infants, probably because their own physical development has not been completed. By age 17 or 18, however, maternal age ceases to be an important risk factor until the mother reaches her late thirties (National Center for Health Statistics 1982). Increasing risks for certain malformations with increasing maternal age suggests that an accumulative environmental factor (or factors) may be working to elevate risks of Down's syndrome and other adverse outcomes (Ayme, Lippman-Hand 1981). One of those factors could be maternal and paternal exposure to low-dose irradiation.

These common maternal factors are important in the epidemiology of fertility and pregnancy outcomes, especially when a woman possesses multiple risks. For example, when maternal ethnic group, age, and parity are considered simultaneously, the ratio of risk for infant mortality between the highest risk groups (a teenaged, black mother or a woman with six or more previous live births) is nine times greater than that of the lowest risk group (nulliparous whites, ages 20-24) (The Collaborative

*Use of trade names is for identification only, and does not constitute endorsement by the Department of Health and Human Services or any of its agencies.

Perinatal Study of the National Institute of Neurological Diseases and Stroke 1972). As a measure of the complexity of these risk factors, let us consider prepregnant maternal weight and two related outcomes--perinatal mortality and infant birth weight. There is a threefold higher risk of low birth weight for the child of the lightest weight mother as compared with the child of the heaviest mother. Conversely, rates of risk for perinatal mortality for low-weight infants increases steadily from mothers weighing the least to those weighing the most with an almost fourfold difference between these two groups (The Collaborative Perinatal Study of the National Institute of Neurological Diseases and Stroke 1972). This increased risk of perinatal mortality with increased maternal weight prior to pregnancy is also found among normal-weight infants.

CONFOUNDING EFFECTS OF COMMON EXPOSURES

The importance of common reproductive risks in the assessment of environmental risks lies in the possibility that risk factors may be confounded in a study of occupational or environmental exposures. Confounding occurs when both the environmental agent and the pregnancy outcome are related to a third (confounding) variable. Problems of confounding variables and statistical methods of dealing with them are discussed by Kleinbaum, Kupper, and Morganstern (1982) and, in a less detailed fashion, by Greenland and Neutra (1980). When one studies a suspected environmental agent, one must consider that common exposures are potential confounding factors, particularly if the distribution of the suspected hazards in the population is correlated with the incidence of known risk factors. For example, paternal, occupational exposure to a chemical may be related to maternal smoking habits because of the joint association of smoking and occupation with socioeconomic status or with some other, correlated factor.

To control for confounding variables, known reproductive risk factors should be included in any study of a potential environmental agent. Then the data should be stratified for each potential confounding factor. Factors which are found to be significantly associated with both the outcome under consideration and the agent being investigated should be controlled for in the analysis of that study.

Hypothetical data presented in Table 2 illustrate how a confounding factor might affect the results of an environmental study. These data could come from a prospective study of 200 exposed men and 400 unexposed men in some occupational setting. In this hypothetical study, the incidence of low birth weight is 6.5 per 100 births for the exposed population and 3.8 per 100 births for the unexposed population, for an observed relative risk of 1.7. This is not statistically significant, however, and the conclusion from such a study could be either that the exposure is not a risk factor or that another study needs to be done to clarify the issue.

Table 2 Hypothetical Low Birth Weight (LBW) Rates for Children of Exposed and Unexposed Male Employees

	N	LBW	Percent LBW
Husband exposed	200	13	6.5
Husband not exposed	400	15	3.8

Relative Risk = 1.7 with 90% confidence interval of 0.95 to 3.2. $\chi_{M-H} = 3.0$, $p = 0.07$

Wives of the men who had been exposed tended to smoke during pregnancy less often than the wives of the unexposed population in this hypothetical study (Table 3). Perhaps they were aware of the potential hazard of their husbands' exposure and therefore selectively reduced their smoking when they discovered they were pregnant. Stratifying for maternal smoking status, therefore, makes a difference in the estimate of relative risk; the adjusted relative risk is 1.9, which is statistically significant. The conclusion now could be that the exposure does elevate risk of low birth weight, controlling for maternal smoking status.

Table 3 Hypothetical Low Birth Weight (LBW) Rates for
Children of Exposed and Unexposed Male
Employees, Controlling for Maternal Smoking

	N	LBW	PERCENT LBW
WIFE SMOKED			
Husband exposed	70	7	10
Husband not exposed	200	10	5
WIFE DID NOT SMOKE			
Husband exposed	130	6	4.6
Husband not exposed	200	5	2.5

Relative Risk = 1.9 with 90% confidence interval of 1.1, 3.5. χ_{M-H} = 1.8, p = 0.04

Data from two studies of cigarette smoking and coffee drinking illustrate that confounding can be important in studies of reproductive outcome (Mau, Netter 1974; van den Berg 1977). In the investigations of coffee drinking and low birth weight, cigarette smoking is a classic confounding factor, since cigarette smoking is related both to coffee drinking and to low birth weight. In these studies, both habits were independently related to the incidence of low birth weight (Table 4). Among heavy smokers, heavy coffee drinkers were about 1.6 times as likely to have an infant weighing less than 2,500 grams (9.9/5.8 = 1.7; 9.0/5.8 = 1.6). When women did not smoke or smoked very little during their pregnancy, their infants were more likely to weigh over 2,500 grams; nevertheless, the heavy coffee drinkers in this group were also more likely to have a low-birth-weight infant (6.6/4.2 = 1.6; 4.4/3.1 = 1.4). The infant at greatest risk (over 9 percent LBW) was one whose mother both smoked and drank coffee excessively, whereas the infant at least risk (between 3 and 4.2 percent LBW) was one whose mother neither smoked nor drank coffee.

Table 4 Relationship of Coffee Drinking and Smoking During Pregnancy to Incidence of Low Birth Weight

| | Percentage Incidence of Low Birth Weight | | | |
| | Heavy Smokers | | Light or Non-Smokers | |
Study	Heavy Coffee Drinkers*	Light or Non-coffee Drinkers	Heavy Coffee Drinkers*	Light or Non-coffee Drinkers
Mau, Netter (1974)	9.9	5.8	6.6	4.2
van den Berg (1977)	9.0	5.8	4.4	3.1

*Average daily consumption exceeded 6 cups per day in van den Berg's study; frequency was not measured in Mau and Netter's investigation.

SYNERGISTIC EFFECT OF RELATED EXPOSURES

When data are confounded, analysis of epidemiologic studies should include testing for both confounding and interaction, since a common exposure may be not only a confounding factor but also a synergistic agent (Kleinbaum, Kupper, Morganstern 1982). Synergism has been found, for example, with lethality of acetaldehyde in combination with nicotine and caffeine in rats. In the doses given, very few deaths occurred when each agent was administered alone. In fact, there were no deaths from the dose of caffeine. Caffeine, in combination with acetaldehyde killed 90 percent of the rats within three hours of dosage. Nicotine and acetaldehyde, in combination killed 50 percent of the rats within three hours and 90 percent within 72 hours of dosing (Sprince, et al. 1981).

Testing for confounding and interaction may reveal groups of persons at particularly high risk of effect for a given environmental agent. We may also gain clues to the mechanisms of action. In this study of rats, for example, the authors found toxic synergy of acetaldehyde with dopamine and protective action of phenoxybenzamine against all synergistic combinations. This suggested to them that excessive catecholamine activity may be involved in the toxic synergy effect.

One type of synergism is peculiar to reproductive epidemiology; an agent that elevates risks of a nonreproductive outcome, such as hypertension or diabetes, may indirectly influence reproductive risks through the effects of the nonreproductive outcome on the mother's ability to have a healthy baby. Thus, subpopulations in the workforce that are more susceptible to those nonreproductive risks should be examined for adverse reproductive outcomes as well.

MEASUREMENT OF COMMON EXPOSURES

To investigate confounding and interaction of common exposures, the exposure history must be measured. Documentation of common exposures may be difficult, however, especially if the investigator relies on maternal recall.

Maternal recall implies knowledge of exposure, ability to recall, and willingness to recall. Knowledge of exposure is a particular problem if one parent is asked to recall the exposure of the other parent. A few studies have investigated the issue of reliability of recall. In one study designed to examine the relationship between pregnancy outcome and drugs taken in early pregnancy for acute illness, recall after delivery tended to be poorer than recall earlier in pregnancy (Klemetti, Saxen 1967).

A traditional case-control study is sensitive to spuriously elevated risks because of selective recall. To avoid this bias, some investigators use the "case-case" approach, with so-called controls being cases hypothesized not to be affected by the potential hazard. If cases are compared with cases also affected by the exposure in an unsuspected way, however, results could be misleading. Since we do not always have a good understanding of potential mechanisms of action, we could be missing important, general effects, or specific effects not previously hypothesized. Even if toxic effects are specific, however, the case-case approach could underestimate the risk if the so-called control group is not representative of the population at risk from which the cases arose. That problem is a likely bias in such studies.

Another approach avoids the problem of representation by selecting a group of pregnant women from the population at risk giving rise to the cases. Women are measured prospectively, during their pregnancy. This may be called a "case-exposure" study since the population at risk sample is measuring the prevalence of exposure in the relevant population. A case-exposure study has similar precision for measurement as a traditional case-control study, plus the added advantages of direct estimate of the relative risk, reduction of selective recall, and a potential measure of effect provided by selective recall (Hogue, Gaylor, Schulz 1983).

SAMPLE SIZE CONSIDERATIONS

Common exposures may themselves be the subject of research. When they are, they are uncommonly difficult to study because large samples must be employed to study modest effect levels. For example, an exposure with 40 percent prevalence would require a case-control study of over 500 cases and 500 controls to detect a 50 percent increased risk with 90 percent power. A cohort study of something like low birth weight with a baseline risk of 5 percent would need to be of a much larger size--over 2,000 in both the exposed and unexposed populations (Interagency Epidemiology Working Group 1980).

With respect to the effects of coffee drinking and low birth weight, 1 of the 2 studies that indicated a 50 percent elevation in low birth weight for women drinking 7 or more cups of coffee daily included 1,100 such women (van den Berg 1977). Interestingly, a recent report entitled, "No Association Between Coffee Consumption and Adverse Outcomes of Pregnancy" (Linn, et al. 1982) involved only 104 women who drank 7 or more cups of coffee a day. The point estimate of relative risk for coffee consumption among smokers, who comprised 76 of the 104 heavy coffee drinkers in this study, was 1.45. The two studies which have reported a significantly elevated risk of low birth weight have had point estimates of 1.53 and 1.58; the power of the Linn study to detect a 50 percent increase in risk is slightly less than 0.5, assuming perfect measurement of coffee consumption. Quite obviously, it is important to consider sample size relevant to the issue of interest.

Another sample-size example illustrates one potential effect on sample size that a confounding factor may have. In the hypothetical illustration of male occupational exposure and maternal smoking (Table 2), there is a benefit derived from controlling for a confounding factor (Table 3). The original sample has little power to detect a relative risk of 2 or less. Yet the stratified relative risk that we found of 1.9 was significant. This illustrates that if there is confounding in a study, controlling for it may, in some instances, increase the power of the study.

In summary, common exposures produce complexity in reproductive epidemiology, but they also may help us to unravel mysteries surrounding less common exposures of interest. If we recognize the need to include their measurement in our studies and if we take advantage of current methodologic approaches, both in design and in analysis of epidemiologic research, we may tease out the effects of less common exposures from the noisy background of more common risk factors.

REFERENCES

Ayme S, Lippman-Hand A (1981). Relations of the maternal age effect in Down syndrome to altered embryonic selection. Am J Epidemiol 144:437.
Bongaarts J (1980). Malnutrition and fecundity. Stud Fam Plann 11:401.
Flynn A, Martier SS, Sokol RJ, Miller SI, Golden NL, del Villano BC (1981). Zinc status of pregnant women: A determinant of fetal outcome. Lancet I:572.
Gal I, Sharman IM, Pryse-Davies J (1972). Vitamin A in relation to human congenital malformations. In Woollam DHM (ed): "Advances in Teratology, Volume 5," New York: Academic Press, p 143.

Greenland S, Neutra R (1980). Control of confounding in the assessment of medical techonology. Int J Epidemiol 9:361.

Hogue CJ (1981). Coffee in pregnancy. Letter to the Editor. Lancet I:554.

Hogue CJR, Gaylor DW, Schulz KF (1983). Estimators of relative risk for case-control studies. Am J Epidemiol 118 (forthcoming).

Interagency Epidemiology Working Group, Food and Drug Administration (1980). An assessment of human epidemiological data concerned with caffeine consumption and problems of pregnancy. Report to the Commissioner, FDA.

Kleinbaum DG, Kupper LL, Morganstern H (1982). Epidemiologic research. Blemont, California: Lifetime Learning Publications.

Klemetti A, Saxen L (1967). Prospective versus retrospective approach in the search for environmental causes of malformations. Am J Public Health 57:2071.

Laurence KM, James N, Miller MH, Tennant GB, Campbell H (1981). Double-blind randomized control trial of folate treatment before conception to prevent recurrence of neural tube defects. Br Med J 282:1509.

Linn S, Schoenbaum SC, Monson RR, Rosner B, Stubblefield PG, Ryan KJ (1982). No association between coffee consumption and adverse outcomes of pregnancy. New Eng J Med 306:141.

Mau G, Netter P (1974). Are coffee and alcohol consumption risk factors in pregnancy? Geburtsh Frauenheilk 34:1018.

National Center for Health Statistics (1982). Advance report of final natality statistics, 1980. Monthly Vital Stat Rept 31(8) Suppl.

Nelson NM (1960). Teratogenic effects of pteroylglutamic acid deficiency in the rat. Proceedings, Ciba Foundation Sumposium.

Rush D, Stein Z, Susser M (1980). A randomized controlled trial of prenatal nutritional supplementation. Pediatrics 65:683.

Shapiro S, Schlesinger ER, Nesbitt REL, Jr (1968). Infant, perinatal, maternal, and childhood mortality in the United States. Cambridge, Massachusetts: Harvard University Press.

Sprince H, Parker CM, Smith GC (1981). Lethal synergy of acetaldehyde with nicotine, caffeine, or dopamine in rats: Protection by ascorbic acid, cysteine, and anti-adrenergic agents. Nutrition Reports International 23(1):43.

Stein ZA, Susser MW (1975). The Dutch famine 1944/45 and the reproductive process. I. Effects on sex indices at birth. Pediatr Res 9(11):70.

The Collaborative Perinatal Study of the National Institute of Neurological Diseases and Stroke (1972). The women and their pregnancies. Washington, D.C.: National Institutes of Health (DHEW Pub. No. (NIH) 73-379).

Thiersch JB (1963). Therapeutic abortions with folic acid antagonist, 4-aminopteroylglutamic acid (4 amino PGA) administered by oral route. Am J Obstet Gynecol 63:1298.

Turner TB, Mezey F, Kimball AW (1977). Measurement of alcohol-related effects in man: Chronic effects in relation to levels of alcohol consumption. Johns Hopkins Med J 141:235.

van den Berg BJ (1977). Epidemiologic observations of prematurity: Effects of tobacco, coffee and alcohol. In Reed DM, Stanley FJ (eds): "The Epidemiology of Prematurity," Munich: Urban & Schwarzenberg, p 157.

FERTILITY AS A MEASUREMENT IN
REPRODUCTIVE TOXICOLOGY

Anthony R. Scialli, M.D.*
Sergio E. Fabro, M.D., Ph.D.
Reproductive Toxicology Center
Columbia Hospital for Women
2425 L Street, N.W.
Washington, D.C. 20037

There is growing concern about the reproductive consequences of exposing human beings to the current large and complex array of environmental and occupational chemicals. Although public awareness has centered on teratology issues, industrial scientists and toxicologists are interested in chemical effects on other aspects of reproduction. It is known that fertility may be dramatically altered by therapeutically administered chemicals (such as oral contraceptives). It is important to consider how environmental and particularly occupational exposures can affect fertility. To detect such effects it is of prime importance to develop sensitive and practical methods by which putative alterations in fertility may be investigated.

INFERTILITY AS A HEALTH PROBLEM

Ten to 15% of couples in the United States are infertile (Coulam 1982). These people have attempted pregnancy for one year without a resulting conception. Infertility is, however, a complex disorder encompassing diverse abnormalities of the male and female genital tracts (see Table). Any attempt to use childlessness as an endpoint in toxicology research must take into account the high background rate of this complaint in the population and the number of different factors which may be involved in failure to conceive.

© **1984 Alan R. Liss, Inc.**

Table 1. Causes of Infertility
(adapted from Coulam 1982)

Cause	Incidence in Infertile Population
Male factors includes decreased sperm number, function, or survival	40%
Obstruction or dysfunction of the fallopian tubes	20-30%
Failure of ovulation	10-15%
Cervical factors includes mucus resistant to sperm penetration or hostile to sperm survival	5%
Unknown	10-20%

The assessment of baseline infertility rates is further complicated by regional differences in populations, by the effects of air temperature and hence seasons on conception rates, and by the influence of age on fertility. Sexually active women between the ages of 16 and 20 have a 95.5% chance of being pregnant after a year of unprotected coitus while between the ages of 35 and 40 the conception rate falls to 68.7% and after age 40 to 30% (Novak, Jones, Jones 1975). Since fertility depends not only on physical and biochemical factors but on behavioral parameters such as coital frequency and technique it is likely that attempts to quantitate infertility as a health problem will always be subject to poorly-defined variables.

METHODOLOGIES IN FERTILITY STUDY

There are two means by which chemical effects on fertility may be measured. The first is a comparison of fertility rates between a group exposed to an agent (or an environment) and a control population. The second involves examination of an exposed group for specific parameters (such as decreased sperm density) known to be important in infertility.

Comparisons of Fertility Rates

Current regulations in many nations provide for the testing of a proposed new drug by administration of the agent to test animals (usually rats or mice) with a comparison of the reproductive performance in exposed and non-exposed groups. To assess the influence of the drug on fertility, some dams are sacrificed in mid-pregnancy for the counting of corpora lutea and resorbing embryos. This is intended as a means to determine what part of any reduced pregnancy rate in the treated group is due to impaired fertility as opposed to early embryolethality. The magnitude of the drug effect on fertility must be very large to produce detectable differences in the test systems used, typically ten male and 20 female animals. Coincident infertility or subfertility of one animal, especially a male, may have a profound effect in test groups of this size (Palmer 1981).

The use of such animal systems is likely to be helpful in examining agents with all-or-none effects on fertility. Antigonadotropins, for example, would be expected to produce easily interpretable results which could be extrapolated to human populations with little difficulty. Most agents, however, would be expected to produce subtle if any changes. Testing of these would require a sensitive measuring tool and must be applicable to humans.

The use of a questionnaire to determine fertility changes in a population of agent-exposed workers has been advocated (Levine et al. 1980). Men occupationally exposed to a nematocide, dibromochloropropane (DBCP), may have reduced sperm counts but the distribution of counts is wide (Milby, Whorton 1980). The suspected effect of DBCP exposure on fertility has been tested by comparing a questionnaire-derived fertility rate with an expected rate (Levine et al. 1981). This method is simple, economical, and requires minimal cooperation on the part of study subjects.

The difficulty with fertility rate comparisons is in choosing an appropriate control population. In the DBCP study expected fertility was derived from US National Birth Probabilities. Study populations, expecially in occupational settings, may be homogeneous in regard to factors known to influence fertility (for example, geographic region, air temperature exposure, and age) while national figures are obtained from heterogeneous groups. Similarly, fertility in a study may be measured restrictively as in counting pregnancies in a population of "ever-married" women. National figures are usually calculated as a number of pregnancies in a population of all reproductive-age women, typically women aged 15 through 44 years (Pritchard, MacDonald 1980). Thus a national fertility rate may serve as a poor control for a local study. Efforts have been made to correct national rates with a constant reflecting local differences (Levine et al. 1981). This constant may be derived from historical data (reproductive performance of workers prior to being hired in the exposed environment) or from examining local groups not exposed to the agent in question (such as the secretatrial staff in the subject workplace).

Fertility is a general designation for a variety of factors and fertility rates in small groups may be greatly affected by unmeasured variables. This is most commonly seen in investigations of exposed workers where it is assumed that all workers have spouses with equivalent reproductive potential. As in the animal test systems discussed above, coincident infertility of one spouse may markedly alter results in a small test population.

An idealized measure of fertility is "fecundability", a mathematical expression of the chances of pregnancy occurring in a given population in a given time period. This can be formulated for population groups where the average "time of waiting" for a conception is known. The reciprocal of the average number of months required for conception is the monthly fecundability, the chance of pregnancy occurring in any one month. It can be shown (Cramer, Walker, Schiff 1979) that the cumulative probability of conception (F) in n months is related to the monthly fecundability (f) by the equation

$$F = 1 - (1-f)^n$$

This yields a life-table type curve when the probability of conception is plotted against time. The curve can be applied for specific populations over short time intervals with great accuracy. If the baseline fecundability of a group can be reliably calculated the effects of exposure to putative fertility-altering agents can be determined with precision.

Measurement of Individual Fertility Parameters

The need to quantitate fertility in a population can be avoided by measurement of isolated factors known to be important in reproductive success. The determination of sperm density and the counting of corpora lutea are reasonably easy and accurate measurements in animals. Animal studies seek to answer three questions about a test chemical:

1. Potential: Can the agent produce reproductive toxicity? This is a yes or no question answered by administering large doses of the test chemical.
2. Potency: What dose is effective for this? This searches for a threshold dose; in most toxicology studies such a threshold is observed (In Utero Alcohol Exposure: Threshold for Effects? 1982).
3. Hazard: Are effective doses less than lethal doses? This provides a basis for comparison between chemicals by determining how close

the effective fertility-altering dose is to the dose that kills a selected percentage of animals.

Animal work is useful in characterizing major effects of fertility- altering substances but has important shortcomings. The application of animal data to humans has been discussed earlier in this symposium. Animal experimentation is also limited in that it frequently deals with the administration of a pure chemical, often by a parenteral route, as a means to gain information about human exposure. Human exposure, however, is to the chemical, its by-products, and contaminants by an occupational or environmental route that may include inhalation, skin absorption, or food contamination.

In formulating human studies identification of the individual fertility parameters likely to be affected by a chemical is essential. In men measurements often include sperm counts (Milby, Whorton 1980) but sperm numbers or densities may not accurately reflect sperm function which includes motility, ovum penetration, and pronucleus formation (Levine et al. 1980). Assessing the quality of the ejaculate with a battery of quantitative measurements may be more helpful (Meyer 1981). Male reproductive potential may, however, be affected by aberrations of libido, erection, or ejaculatory function. These are more difficult to measure and often require reporting through interviews or questionnaires which may be biased by subject embarrassment and denial.

Women can be evaluated for the regular occurrence of ovulation. This is detectable by recording the daily body temperature on awakening. Progesterone production after ovulation causes a 0.5-1.0 $^{\circ}F$ rise in basal temperature. Endometrial biopsy or peripheral blood sampling can also be used to detect ovulation. Ovulation is an all-or-none phenomenon and as such is insensitive as a measure of subtle chemical effects. Poor corpus luteum function may be responsible for infertility even in the presence of seemingly-normal ovulation (American College of Obstetricians and Gynecologists 1981). Histological evidence of normal recent ovulation may be found in women after the age of 50 when conception is highly unlikely (Novak, Jones, Jones 1975).

Of increasing use in both men and women are biochemical tests thought to bear on reproductive potential. Substances such as follicle stimulating hormone, leutinizing hormone, prolactin, sex steroid binding globulin, and testosterone have a reasonably narrow range of normal values and have been used to assess the fertility effects of occupational exposure to chemicals (Willems 1981). Although exact measurements and comparisons of blood hormone levels may be made, it is unclear whether modest alterations in these levels have significant implications for reproductive potential.

CONCLUSION

The use of fertility as an endpoint in toxicological research is hampered by the large number of variables influencing human reproductive potential. Comparisons of chemically exposed and non-exposed population fertility rates require careful selection of control groups or knowledge of the baseline (expected) fecundability of the study group. Individual parameters believed to be important in fertility may be more readily measured; however, except when a fertility factor is severely compromised, data on individual parameters must be interpreted with caution. Subtle alterations of a single factor in the complex mechanism of fertility may not be significant against the high background rate of reproductive failure in human beings.

American College of Obstetricians and Gynecologists (1981). Luteal phase defect. In Stone ML (ed): "ACOG Update", 12:1-9.
Coulam CB (1982). The diagnosis and management of infertility. In Sciarra JJ (ed): "Gynecology and Obstetrics", New York: Harper and Row, p 1-18.
Cramer DW, Walker AM, Schiff I (1979). Statistical methods in evaluating the outcome of infertility therapy. Fert Steril 32:80-86.
In Utero Alcohol Exposure: Threshold for Effects?

(1982). In Fabro S (ed): "Reproductive Toxicology, A Medical Letter", Washington, DC: Reproductive Toxicology Center, 1:11-14.

Levine RJ, Symons MJ, Balogh SA, Arndt DM, Kaswandik NT, Gentile JW (1980). A method for monitoring the fertility of workers. 1. Method and pilot studies. J Occup Med 22:781-791.

Levine RJ, Symons MJ, Balogh SA, Milby TH, Whorton MD (1981). A method for monitoring the fertility of workers. 2. Validation of the method among workers exposed to dibromochloropropane. J Occup Med 23:183-188.

Meyer CR (1981). Semen quality in workers exposed to carbon disulfide compared to a control group from the same plant. J Occup Med 23:435-439.

Milby TH, Whorton D (1980). Epidemiological assessment of occupationally related chemically induced sperm count suppression. J Occup Med 22:77-82.

Novak ER, Jones GS, Jones HW (1975). "Novak's Textbook of "Gynecology." Baltimore: Williams & Wilkins Co., pp 625-626.

Palmer AK (1981). Regulatory requirements for reproductive toxicology: theory and practice. In Kimmel CA, Buelke-Sam J (eds): "Developmental Toxicology," New York: Raven-Press, pp259-287.

Pritchard JA, MacDonald PC (1980). "Williams Obstetrics." New York: Appleton-Century-Crofts, p 2.

Willems H (1981). Occupational exposure to estrogens and screening for health effects. J Occup Med 23:813-817.

LABORATORY TESTS FOR HUMAN MALE REPRODUCTIVE RISK ASSESSMENT

James W. Overstreet, M.D., Ph.D.

School of Medicine
University of California, Davis
Davis, California   95616

INTRODUCTION

When the output of spermatozoa from the testes is compared among various species, it is apparent that human sperm output is about four times less than that of other mammals in terms of the number of sperm produced per gram of tissue (Amann and Howards, 1980; Johnson et al., 1980). The human ejaculate is also virtually unique in the number of abnormal sperm cells which are present in the ejaculate of a normal male (MacLeod and Gold, 1953). We do not even consider a man to have a potential fertility problem until the proportion of abnormal cells in the semen exceeds fifty per cent. Human males could therefore be more vulnerable to environmental effects than are other mammals. Our methods of risk assessment, which principally involve studies of laboratory animals, may underestimate the impact of environmental hazards on human male fertility. There is a growing awareness that potential reproductive hazards must be studied in exposed men as well as animal models. The purpose of this chapter is to describe the laboratory and clinical tests now being developed for assessment of reproductive risks to human males from environmental hazards.

SELECTION OF TESTS FOR MALE REPRODUCTIVE RISK ASSESSMENT

The types of studies in which reproductive tests may be applied fall into two general categories. First, there may be surveillance studies of men who are exposed to compounds or environments that may be potentially toxic, but are not

© 1984 Alan R. Liss, Inc.

known to be specific reproductive hazards. Such a situation could occur in the workplace where surveillance studies might involve longitudinal, long-term assessments, perhaps as part of a yearly physical examination. The purpose of such a program is to detect small but significant changes in the reproductive function of exposed men. The second type of study may involve a group of individuals who have been exposed to a suspected reproductive toxicant (e.g., as identified in animal studies). In this case there may be a more limited number of assessments, and the purpose of the study is to determine whether or not the reproductive potential of the exposed men has actually been affected. The types of tests which are appropriate in these two situations are somewhat different and the interpretation of the tests may also be different.

In selecting tests for male reproductive risk assessment there are several important criteria to be considered. The tests must be objective, and they must generate quantitative data which are amenable to statistical analysis. The tests must be technically sound; they must be accurate and precise. The tests should measure a biologically stable parameter, that is, the normal variation should not overshadow a small but significant change. These tests should be sensitive to early warning signs of reproductive toxicity and not simply confirm male sterility in the population. Finally, the tests must be technically feasible, they must be financially feasible and they must be culturally feasible. These criteria will be taken into account in describing the tests which are available for human male reproductive risk assessment.

Tests of male reproductive function fall into three major categories. Most involve some type of semen analysis and this will be the emphasis for discussion in this chapter. However, it is also worthwhile to mention two other types of evaluation, since these pertain to any overall assessment of male reproductive risk. They are the physical examination and the evaluation of endocrine function. These types of tests will also be considered briefly in this chapter.

INTERPRETATION OF REPRODUCTIVE TESTS IN THE ABSENCE OF CLINICAL INFERTILITY

It may be difficult to interpret laboratory tests of reproductive function for men who are not attempting pregnancy.

In general, the "normal" values for these tests have been
determined by comparison of groups of known fertile men with
groups of infertility patients. Virtually all of the clinical
experience with these tests has been with clinically infertile couples (i.e., trying to conceive without success for
at least one year). The limited information which is available suggests that men with proven previous fertility (e.g.,
candidates for vasectomy) may have semen parameters which
would be considered abnormal by the usual clinical criteria
(Nelson and Bunge, 1974; Zuckerman et al., 1974). There are
no data available with which to calculate a true normal
range for any of the laboratory tests of human semen. This
means that studies of reproductive toxicity must be carried
out by comparison with carefully matched, non-exposed control subjects. Only in cases of very severe semen abnormality can any prediction of fertility potential be given
for a specific individual in the absence of clinical infertility.

THE PHYSICAL EXAMINATION OF THE TESTES

The physical examination is of obvious importance in
the assessment of male reproductive function. When the
physician examines the male genitalia, the size and consistency of the testes are recorded, and it is important that
objective and reliable information be obtained. Testis size
should be measured with calipers to determine length and
width. Additional information on testicular volume may be
gained by direct comparison with standard models of known
volume. This is a subjective assessment but it may provide
additional useful information for surveillance studies which
monitor changes in the size of the gonads from year to year.

The consistency of the testes is also evaluated in the
physical examination and objective information on this characteristic can be obtained by using a tonometer. This instrument is the one used by ophthalmologists for assessing
intraocular pressure, and intratesticular pressure can be
measured in an analogous manner. This methodology is currently in the research phase of development. The utility
of testicular tonometry in predicting male fertility has not
been determined, but the method is already being used by our
laboratory in longitudinal surveillance studies to provide
objective data on changes in testicular consistency over
time. This methodology has a unique importance for studies

of populations which include vasectomized men. In these
cases there are no spermatozoa which can be studied for
evidence of testicular toxicity. Nevertheless, the testis
continues to function after vasectomy and it is still an
organ which is susceptible to toxic effects. It is possible
that testicular tonometry will allow us to monitor those
effects in the vasectomized men as well as in those men who
are producing spermatozoa in the semen.

ENDOCRINE EVALUATION

The role of endocrine evaluation in the assessment of
male reproductive toxicity will not be discussed in detail.
A toxic effect on the pituitary or on the central nervous
system could result in low levels of circulating gonadotropins (LH and FSH) and thus affect spermatogenesis. A direct
toxic effect on the testes is more likely, and this could
lead to abnormally high levels of gonadotropins because of
the feedback relationship between the pituitary and the
gonads. Radioimmunoassays for gonadotropins are well established and technically sound. They also meet the criteria
for biological stability. Since these tests are carried
out with blood, they are more feasible from the cultural
standpoint than tests on the semen. It may be difficult to
motivate exposed individuals to provide semen samples for
analysis, whereas the physical examination and analysis of
blood are more likely to be acceptable. However, physical
measurements of the testes and laboratory assays of circulating hormones are likely to provide only late signs of
reproductive toxicity. It is likely that irreversible
damage to the germinal epithelium will have occurred before
endocrine changes or physical alterations are apparent.

SEMEN EVALUATION

In the standard semen evaluation assessment is made of
semen volume, sperm concentration, sperm motility and sperm
morphology. Measurements of sperm numbers (i.e., sperm
concentration and total sperm numbers) are the most feasible
component of the semen evaluation. The semen samples can be
collected at a distant location and no special conditions
are required for shipment to the base laboratory. The methods
for counting sperm are simple and well established. Until
recently, only the sperm concentration and the ejaculate

volume were parameters of the semen which could be measured
objectively. The need to rely on subjective assessment of
sperm motility and sperm morphology has greatly limited the
value of the semen evaluation.

Sperm Concentration

Our assumptions about the number of spermatozoa which
are adequate for fertility derive from the classical studies
of John MacLeod, which were carried out in the 1940s.
MacLeod compared the sperm concentration (millions of sperm
per ml of semen) in a group of 1,000 infertile men with a
group of 1,000 men whose wifes were pregnant at the time of
the evaluation. He found no difference between the groups
except for the population of men with sperm counts less than
20 million sperm per ml. These men were encountered more
frequently in the infertile group than in the fertile group
(MacLeod and Gold, 1951). Today this remains the most widely
accepted value for the lower limit of a "normal" sperm count.
It has recently been speculated that sperm counts of American
men may have declined during the last 40 years and that the
normal parameters derived from the studies of MacLeod are no
longer valid (Nelson and Bunge, 1974; Zuckerman et al., 1977).
However, no such trend has been detected during this time in
the clinic population evaluated by MacLeod's laboratory
(MacLeod and Wang, 1979). Therefore, we have no reason to
change our notion of the normal sperm concentration nor should
we be overly concerned that there has been a precipitous
change in sperm output of American men in the last 4 decades.
The data on sperm counts from MacLeod's laboratory have revealed a remarkable consistency in the proportion of infertility patients with sperm counts lower than 20 million
sperm per ml; this has remained somewhat less than 20% of
the total population (MacLeod and Wang, 1979). Thus, more
than 80% of these infertility patients have sperm counts
which meet or exceed the normal standards. Obviously, there
are other significant abnormalities in the semen of these
men and assessment of sperm count alone would not identify
them as having a fertility problem.

The use of sperm counts for reproductive risk assessment is also confounded by biological instability of this
parameter in humans. Our laboratory has studied repeated
semen samples from fertile donors over a period of one year
(Katz et al., 1982a). Coefficients of variation in these

samples for sperm concentration and total sperm number per ejaculate were 0.46 and 0.58, respectively. Thus, even in normal men there is very significant day to day variation in the sperm count. This normal variation makes it very difficult to assess subtle changes in testicular function on the basis of sperm count alone. The other significant disadvantage of using sperm counts for reproductive risk assessment is that they may reflect relatively late signs of testicular toxicity. By the time conclusive information is obtained suggesting sperm count depression in a group of men, it is very possible that irreversible testicular damage and sterility may have occurred in some members of that population.

Sperm Motility

There is information in the semen evaluation which could provide earlier warning signs of reproductive dysfunction in exposed men. It has long been appreciated by clinicians and basic scientists that the motility of the sperm cells in the semen should be an important indicator of a man's fertility. The movement characteristics of spermatozoa reflect many of the internal mechanisms within the cell which are required for its transport through the reproductive tract and for penetration of the ovum (Overstreet and Katz, 1981). In the past, the value of this parameter has been limited because sperm motility assessment was subjective. The percentage of moving cells usually has been estimated rather than counted and the quality of movement has been rated on an arbitrary scale of 0 to 4+ (Eliasson, 1975). A number of methods are now available which generate objective data on sperm movement. These methods measure the percentage of sperm cells that are moving as well as their swimming speed. Some methods can provide information on the swimming trajectory (how straight the sperm swim) as well as on the frequency and shape of the flagellar waves (Katz and Overstreet, 1979).

The simplist and most practical method for sperm motility assessment involves the use of videomicrography (Katz and Overstreet, 1981). The equipment required includes a microscope, a video camera which is mounted on the microscope, a videotape recorder and a monitor. A videotape recording is made of the spermatozoa swimming in the semen. When the tape is replayed, percentage motility is determined by

marking individual sperm cells on the screen, advancing the
tape and scoring the sperm as motile if any movement is
detected. To measure swimming speeds, a transparent overlay
is used with concentric rings which have been calibrated
using a micrometer. The overlay is applied to the video
screen with its center on the head-midpiece junction of the
spermatozoon being assessed. The tape is advanced for a
known interval of time and the swimming speed is calculated
by a reference to the specific ring which the sperm has
reached. This technique has proven quite repeatable and
provides objective quantitative data (Katz and Overstreet,
1981). The percentage of motile sperm is usually determined
from assessment of 50 to 100 cells distributed in several
fields of the microscope slide. The mean swimming speed is
usually based on measurement of 25 to 50 cells. In our
study of repeated semen samples from fertile donors, the
coefficients of variation for percent motility and mean swimming speed were 0.27 and 0.19, respectively (Katz et al.,
1982a). Thus, normal biological variation in these parameters was substantially less than that encountered when sperm
numbers were assessed. The principal disadvantage of sperm
motility measurements for human reproductive risk assessment
is one of feasibility. Artifacts of sperm motility can
easily be induced during collection or transport of the
semen. Video recording must be made within one to two hours
of semen collection. These difficulties may be minimal when
longitudinal studies are carried out in established laboratories. However, these problems may be more difficult to
overcome in field studies of limited duration.

Sperm Morphology

Human semen typically contains many abnormal sperm cells.
The abnormalities which may be observed include abnormalities
of sperm size and shape. The sperm head, midpiece or tail
may be affected. There are numerous reports in the clinical
literature which associate these abnormal sperm shapes with
male fertility problems (MacLeod, 1971). There is good
reason to believe that morphological changes in the seminal
sperm may be a sensitive indicator of acute stress to the
testes (MacLeod, 1974). Unfortunately, there has been no
objective methodology for assessing sperm morphology. There
are no authoritative atlases of human sperm morphology for
teaching or for reference.

Videomicrography has also proven useful as a tool for objective assessment of sperm morphology (Overstreet et al., 1981). The sperm are assessed on stained seminal smears. Each sperm is viewed on the video monitor at a total magnification of approximately 5,000 diameters. The cells can be recorded on videotape as a permanent record, or the sperm image may simply be projected on the video screen without recording. The morphology of each sperm is assessed on the videoscreen using a transparent overlay analogous to that employed in motility assessment (Katz et al., 1982b). The overlay is calibrated with metric standards for sperm head length and width. Fifty sperm are evaluated on each slide. In our study of day to day variation in semen quality, the coefficient of variation for morphologically normal sperm was only 0.15 (Katz et al., 1982a). Thus, sperm morphology appears to show the least biological variation of any of the standard semen parameters. Sperm morphology assessment is very feasible for field studies since seminal smears can be made when the ejaculate is obtained, and the slides may be stained and analyzed subsequently at the base laboratory.

Even with our current methods of objective assessment, the value of sperm morphology evaluation has not been fully realized. In part this is because the human element has not been eliminated from the decision making process. Another limiting factor is our expression of the sperm morphology parameter as a percentage of "normal" sperm. Although the normal sperm is specified by metric standards in our laboratory, there are no sound biological or clinical data on which such standards can be based, and there is no agreement among clinicians as to how these standards should be set (Freund, 1966).

The problems associated with classifying sperm by type can be overcome by directly assessing the morphometry of the sperm cells. The sperm image on the video screen can be digitized and the information can be fed directly into a small computer for calculation of sperm head length, width, circumference, area, etc. (Katz et al., 1982a). We are currently pursuing the idea that a single parameter of sperm morphometry could be calculated for application to human reproductive risk assessment. The approach is based on the observation that spermatozoa in high quality human semen are fairly uniform in size and shape, whereas abnormal semen contains a diversity of head sizes and shapes (MacLeod, 1964). Furthermore, the magnitude of this diversity tends to be proportional to the degree of abnormality in the semen quality. Our preliminary experiments indicate that this diversity can be objectively measured by calculating the ratio of head length to head width for 50 sperm cells and a standard deviation for this parameter. Further studies with large populations will be required to establish a "normal range" for such a new parameter. However, the method is now useful for comparison of suitably selected exposed and control populations.

TESTS OF SPERM CERVICAL MUCUS INTERACTION

As the spermatozoa travel to the site of fertilization, they must pass first through the cervix and its complement of cervical mucus (Overstreet and Katz, 1981). The cervical mucus is a well defined fluid which is accessable for study in the laboratory. Important functions of the sperm cell can be assessed by examination of sperm interaction with the cervical mucus (Moghissi, 1976). These functions are related to the sperm flagellar activity and to the cell surface, since the sperm must interact very closely with the microstructure of the cervical mucus. For example, the presence of anti-sperm antibodies on the sperm surface may be first suspected when there is an abnormality of sperm cervical mucus interaction (Jager et al., 1979).

Sperm cervical mucus interaction is assessed either on a slide or in a capillary tube. In the slide test, semen and cervical mucus are placed in contact and observations are made of the ability of the sperm cells to enter into the mucus and of their swimming behavior in the cervical mucus. Application of this test to reproductive risk assessment is limited by its lack of objectivity. Capillary tube systems can provide more quantitative information (Kremer, 1965). In these tests, mucus is drawn into the capillary tube and the tube is then exposed to semen. The results of the test can be expressed as the distance traveled by spermatozoa along the tube during a prescribed period of time. When optically clear, flat capillary tubes are used, it is also possible to make observations on the motility and number of sperm at various locations along the tube (Katz et al., 1980). Recently, we have developed a method for obtaining objective data on sperm penetration of the mucus. The concentration of motile sperm in the semen and their mean swimming speed are determined, and from this information the expected number of collisions between the sperm and the mucus interface can be calculated. The number of sperm which enter the mucus is then counted and compared with the number of predicted collisions. From the ratio of these numbers a percentage of successful collisions (PSC) can be calculated (Katz et al., 1980). When flat capillary tubes are used, we can also measure the swimming speed of sperm in the mucus as we do in semen.

Although cervical mucus penetration tests are useful clinically, there are feasibility problems in applying this kind of test to assessment of reproductive hazards. Human cervical mucus is very difficult to obtain in quantity. The human female produces relatively small volumes of mucus and the mucus is receptive to sperm for only a few days during the menstrual cycle. There is a possibility that bovine cervical mucus, which is available in large quantities and is easily obtained, can be substituted for human mucus (Moghissi et al., 1982). Research is also in progress to develop synthetic simulants for cervical mucus (Lorton et al., 1981). If these substitute systems are proven feasible, they may have a role in reproductive risk assessment.

TESTS OF SPERM OOCYTE INTERACTION

A great deal of scientific information is now available on the biology of fertilization in mammals (Yanagimachi, 1981). It is believed that the human sperm cell must undergo a number of physiological changes prior to fertilization and these are collectively known as capacitation. No visible change in the spermatozoon is apparent after completion of capacitation. The next step in the fertilization process, the acrosome reaction, is a morphological change which can be demonstrated in the laboratory (either specialized cytology techniques or electron microscopy is required). However, these methods are too laborious and technically complex for routine application to reproductive risk assessment (Overstreet, 1983).

For ethical reasons, we cannot attempt to fertilize human eggs in vitro for diagnostic purposes. However, there are substitutes which can be used for the human egg to monitor some of the functions of the sperm cell which are involved in fertilization. A number of years ago it was determined that the egg of the golden hamster has its primary block to interspecies fertilization at the level of the zona pellucida (Yanagimachi et al., 1976). The zona pellucida is an acellular investment of the oocyte which can be easily removed in the laboratory by enzymic digestion. Once the zona pellucida of the hamster egg is removed, spermatozoa of many species including humans can fuse with the egg (Yanagimachi, 1981).

Both capacitation and the acrosome reaction are required for human sperm fusion with the zona-free hamster egg, and the hamster egg test is therefore an appropriate assay for these sperm functions (Overstreet, 1983). The conditions for this laboratory test have not been standardized. In our laboratory, the semen is incubated under a layer of culture medium for a period of one hour. During this time, the sperm swim out of the semen and into the culture media. The sperm suspension is centrifuged and the sperm pellet is then washed several times to obtain a cell suspension which is free of seminal plasma. The sperm suspension is then incubated to achieve capacitation. The capacitated sperm are mixed in a petri dish with the zona-free hamster eggs and several hours are allowed for sperm penetration to occur. The test results are expressed as the percentage of eggs penetrated and the number of penetrating sperm in each egg.

There are a number of clinical reports of a close association between male fertility and the success of sperm fusion with zona-free hamster eggs in vitro (Rogers et al., 1979; Karp et al., 1981). There is controversy concerning the normal value for this test. Some laboratories consider that when fewer than 15% of the hamster eggs are penetrated there is evidence of male infertility. In our laboratory, we do not consider the test abnormal if any eggs are penetrated. There appear to be men who have normal semen by all other criteria, but who produce sperm which are unable to fuse with the hamster egg and therefore appear to be infertile. Thus, this test may detect deficiencies in the sperm cells which would not be apparent from other tests of male fertility.

The feasibility of implementing this test is currently limited because it can only be done in a few specialized laboratories. Our laboratory has recently developed a method for transporting semen to such specialized centers for testing. The fresh semen sample is diluted with a balanced salt solution containing egg yolk. The container of diluted semen is then transferred to an ice chest for shipment. We have shown that semen can be stored in such a manner for at least 48 hours (Bolanos et al., 1983). This is ample time for commercial transportation services to deliver semen to a specialized laboratory after its collection in the field.

SUMMARY AND CONCLUSIONS

The criteria for reproductive test selection which were set forth in the beginning of this chapter required that the tests be objective, technically sound, biologically stable, sensitive and feasible. All of the tests which have been discussed can generate objective quantitative data (Table 1). Testicular tonometry appears to be a technically sound procedure which measures a biologically stable parameter, although this remains to be proven. Sperm counts are definitely not a biologically stable parameter. There is insufficient information to judge the biological stability of data obtained from sperm cervical mucus interaction. Data from a number of laboratories suggest that the zona-free hamster egg assay gives stable results when repeated with the same donor, and the tests as performed in specialized laboratories are technically sound at the present time. However, the number of laboratories which can perform the assay is limited.

Sensitivity to early toxicity is a very important criterion for test selection. Physical examination does not meet this criterion, endocrine studies do not, and sperm counts do not. Not enough information is currently available to determine the sensitivity of sperm motility assessment. Sperm morphology assessment may be the most sensitive early indicator of reproductive toxicity which is currently available. There is a large body of clinical and basic science literature which suggests that sperm morphology may reflect acute stress effects on the testes.

The feasibility of these tests vary. Sperm motility may be feasible only in longitudinal studies in which the video equipment can be set up in a laboratory which is doing repeated assessments. Sperm morphology assessment does not require any specialized equipment in the field. Studies of sperm cervical mucus interaction, for the reasons already stated, remain non-feasible at this time. Tests of sperm-egg interaction are probably feasible if spermatozoa can be shipped to a

TABLE 1. ASSESSMENT OF CONTEMPORARY TESTS FOR HUMAN MALE REPRODUCTIVE TOXICITY

Criteria	Physical Examination	Endocrine Studies	Sperm Counts	Sperm Motility	Sperm Morphology	Mucus Studies	Oocyte Studies
Objective Data	Yes	Yes	Yes	Yes	Yes	Yes	Yes
Technically Sound	Probably	Yes	Yes	Yes	Yes	Yes	Probably
Biological Stability	Probably	Yes	No	Yes	Yes	?	Yes
Sensitivity to Early Toxicity	No	No	No	?	Possibly	?	?
Feasible	Yes	Yes	Yes	Surveillance Only	Yes	No	Probably

specialized laboratory for assessment. Thus, there are now a number of new tests for male reproductive function which are available, and which are practical. It is time for this technology to be transferred from the basic science laboratory for application in human reproductive risk assessment.

ACKNOWLEDGMENTS

The unpublished studies described in this chapter were supported by grants from the National Institutes of Health (HD15149), the National Institute for Occupational Safety and Health (OH 01148) and the Environmental Protection Agency (R809089). A Research Career Development Award from the National Institutes of Health (HD00224) is also acknowledged.

REFERENCES

Amann RP, Howards SS (1980). Daily spermatozoal production and epididymal spermatozoal reserves of the human male. J Urol 124:211.

Bolanos JR, Overstreet JW, Katz DF (1983). Human sperm penetration of zona-free hamster eggs after storage of the semen for 48 hours at 2°C to 5°C. Fertil. Steril. In press.

Eliasson R (1975). Analysis of semen. In Behrman SJ, Kistner RW (eds): "Progress in Infertility," Boston: Little Brown and Co, p 691.

Freund M (1966). Standards for the rating of human sperm morphology. Int J Fertil 11:97.

Jagar S, Kremer J, van Slochteren-Draaisma T (1979). Presence of sperm agglutinating antibodies in infertile men and inhibition of in vitro sperm penetration into cervical mucus. Int J Androl 2:117.

Johnson L, Petty CS, Neaves WB (1980). A comparative study of daily sperm production and testicular composition in humans and rats. Biol Reprod 22:1233.

Karp LE, Williamson RA, Moore DE, Sky KK, Plymate SR, Smith WD (1981). Sperm penetration assay: useful test in evaluation of male fertility. Obstet Gynecol 57:620.

Katz DF, Overstreet JW (1979). Biophysical aspects of human sperm movement. In Fawcett DW, Bedford JM (eds): "The Spermatozoon," Baltimore: Urban and Scharzenberg, Inc., p 412.

Katz DF, Overstreet JW (1981). Sperm motility assessment by videomicrography. Fertil Steril 35:188.

Katz DF, Diel L, Overstreet JW (1982b). Differences in the movement of morphologically normal and abnormal human seminal spermatozoa. Biol Reprod 26:566.

Katz DF, Overstreet JW, Hanson FW (1980). A new quantitative test for sperm penetration into cervical mucus. Fertil Steril 33:179.

Katz DF, Overstreet JW, Pelfrey RJ (1982a). Integrated assessment of the motility, morphology and morphometry of human spermatozoa. In Spira P, Jouannet P (eds): "Human Fertility Factors," Paris: Inserm, p 97.

Kremer J (1965). A simple sperm penetration test. Int J Fertil 10:209.

Lorton SP, Kummerfeld HL, Foote RH (1981). Polyacrylamide as a substitute for cervical mucus in sperm migration tests. Fertil Steril 35:222.

MacLeod J (1964). Human seminal cytology as a sensitive indicator of the germinal epithelium. Int J Fertil 9:281.

MacLeod J (1971). Human male infertility. Obstet Gynecol Surv 26:335.

MacLeod J (1974). Effects of environmental factors and of antispermatogenic compounds on the human testis as reflected in seminal cytology. In Mancine RE, Mastini L (eds): "Male Fertility and Sterility," New York: Academic Press, p 123.

MacLeod J, Gold RZ (1951). The male factor in fertility and infertility. II. Spermatozoon counts in 1000 men of known fertility and in 1000 cases of infertile marriage. J Urol 66:436.

MacLeod J, Gold RZ (1953). The male factor in fertility and infertility. VI. Semen quality and certain other factors in relation to ease of conception. Fertil Steril 4:10.

MacLeod J, Wang Y (1979). Male fertility potential in terms of semen quality: a review of the past, a study of the present. Fertil Steril 31:103.

Moghissi KS (1976). Postcoital test: physiologic basis, technique and interpretation. Fertil Steril 27:117.

Moghissi KS, Segal S, Meinhold D, Agronow SJ (1982). In vitro sperm cervical mucus penetration: studies in human and bovine cervical mucus. Fertil Steril 37:823.

Nelson CMK, Bunge R (1974). Semen analysis: evidence for changing parameters of male fertility potential. Fertil Steril 25:503.

Overstreet JW (1983). Evaluation and control of the fertilizing power of sperm. In Andre J (ed): "The Sperm Cell,"

Boston: Martinus Nijhoff Publishers, p 1.

Overstreet JW, Katz DF (1981). Sperm transport, capacitation. In Speroff L, Simpson JL (Eds): "Gynecology and Obstetrics, vol. V. Reproductive Endocrinology, Infertility and Genetics," Philadelphia: Harper and Row Publishers, Chapter 45.

Overstreet JW, Price MJ, Blazak WF, Lewis EL, Katz DF (1981). Simultaneous assessment of human sperm motility and morphology by videomicrography. J Urol 126:357.

Rogers BJ, Campen HV, Ueno M, Lambert H, Bronson R, Hale R (1979). Analysis of human spermatozoal fertilizing ability using zona-free ova. Fertil Steril 32:644.

Yanagimachi R (1981). Mechanisms of fertilization in mammals. In Mastroianni L, Jr, Biggers JD (eds): "Fertilization and Embryonic Development In Vitro," New York: Plenum Publishing Corporation, p 81.

Yanagimachi R, Yanagimachi H, Rogers BJ (1976). The use of zona-free animal ova as a test-system for the assessment of the fertilizing capacity of human spermatozoa. Biol Reprod 15:471.

AN EVALUATION OF SPERM TESTS AS INDICATORS OF GERM-CELL DAMAGE IN MEN EXPOSED TO CHEMICAL OR PHYSICAL AGENTS*

A.J. Wyrobek, G. Watchmaker, and L. Gordon

Lawrence Livermore National Laboratory
Biomedical Sciences Division, L-452
University of California, P.O. Box 5507
Livermore, California 94550

I. INTRODUCTION

The semen studies performed on men working with dibromochloropropane (DBCP) provide clear examples that an occupational chemical can have profound anti-spermatogenic effects (Babich et al., 1981). However, DBCP is certainly not the only chemical agent known to affect human sperm production. As reviewed here, at least 89 chemical exposures have been studied for their effects on human spermatogenesis using sperm tests, with the majority showing some effect on sperm count, motility, or morphology. Approximately 85% of these exposures were to experimental or therapeutic drugs, 10% to occupational or environmental agents, and 5% to recreational drugs (Wyrobek et al., 1983a).

A number of non-sperm approaches have been used to assess induced changes in testicular function in man: testicular biopsy, questionnaire survey, and blood levels of gonadotrophins. Testicular biopsy can provide a direct cytological assessment of spermatogenesis in situ (Whorton et al., 1979). However, it is a surgical procedure with associated trauma and cost, and may itself affect testicular function. Epidemiological surveys of

---

*Also printed in Proceedings of the 5th Annual Occupational and Environmental health Conference, held in Park City, Utah, April 1983.

Work performed under the auspices of the U.S. Department of Energy by the Lawrence Livermore National Laboratory under contract number W-7405-ENG-48.

© 1984 Alan R. Liss, Inc.

reproductive function using questionnaires have been shown to be effective for identifying male-linked infertility (Levine et al., 1981) and increased rates of spontaneous abortion (Morgan et al., 1983). Epidemiological surveys, however, generally require large sample sizes and are expensive. Analyses of blood levels of gonadotrophins are also expensive, and are generally insensitive to small changes in spermatogenic function (Whorton et al., 1977; Whorton et al., 1979). Though these approaches are useful in certain settings, more sensitive and direct methods for assessing human male reproductive toxicity are needed.

Sperm tests have a long history in the diagnosis of infertility in man and domestic animals. It was therefore not surprising that early attempts to assess chemically altered human spermatogenic function used sperm parameters common to fertility diagnosis, i.e., sperm number (counts), motility, and morphology (seminal cytology). Many animal and human studies have shown that sperm anomalies can be used as indicators, and, in certain cases, dosimeters of induced antispermatogenic effects (for reviews see Wyrobek et al., 1983a,b). Though induced sperm anomalies are clearly linked to testicular damage, their relationship to reduced fertility and to induced, heritable, genetic defects is not yet clear and is an active research area. Sperm tests have received much attention from those who wish to monitor human exposure to germ cell mutagens and male reproductive toxins, because (a) sperm are the only human germ cell that can be easily obtained, (b) they can reflect damage to the gonads, and (c) they can be studied both in humans and animal models.

This paper briefly describes the more common sperm-based methods and reviews some of their applications. It also includes guidelines for undertaking a human sperm study, as well as a discussion of the predictive value of induced sperm changes, an evaluation of the role of animal sperm tests, and a summary of the advantages and disadvantages of the sperm tests.

II. HUMAN SPERM TEST METHODS

A. Sperm Counts

Sperm count is reported as the number of sperm per milliliter of ejaculate (or as the total number of sperm ejaculated), and is usually determined by hemocytometer. The measurement is technically easy and automated methods are available. Sperm count has been the single most commonly employed test to assess the effects of physical and chemical agents on human spermatogenesis (for a review of radiation effects, see Ash, 1980, and for a review of chemical effects, see Wyrobek et al., 1983a). Interpretation of results may be confounded by a number of factors, however, such as variable continence time prior to sample collection, frequency of sexual contact, and collection of incomplete ejaculate (Schwartz et al., 1979). The measure is variable even in normal unexposed men, and there has been considerable controversy as to what constitutes a normal sperm count in relation to fertility. Though the underlying variability in this measure leads to statistical problems in detecting small induced changes, there are many clear examples of agent-induced reductions in sperm counts (Wyrobek et al., 1983a).

B. Sperm Motility

Sperm motility is the swimming ability of the sperm, and has been expressed in a variety of ways, such as percentage of motile sperm, or on a graduated scale of 1 to 4. Some laboratories also distinguish between progressive and nonprogressive sperm movement. Although motility may be one of the better performance evaluations of spermatogenic function in relation to fertility, it is also the most subjective parameter of the four discussed in this review, and is very sensitive to time and temperature after sample collection (Makler et al., 1979). Thus semen motility is a difficult if not impossible parameter to measure in a large-scale cross-sectional study, especially when samples are collected at home. Normal ranges of sperm motility have been difficult to define and are usually laboratory- and scorer-specific (Amelar, 1966). Much effort has gone into the development of automated methods, and several approaches are promising (e.g., Katz and Overstreet, 1981; Jouannet et al., 1977).

C. Sperm Morphology

Sperm morphology (also referred to as seminal cytology) is the visual assessment of the shapes of sperm in an ejaculate. Although sperm-head shape is usually emphasized, some assessments also incorporate midpiece and tail abnormalities as well as immature forms. The visual assessment of sperm morphology is very subjective, and is critically dependent on the classification scheme used. For the highly angular shapes of the sperm heads of laboratory rodents it is relatively easy to detect visually subtle deviations in shape. In contrast, the more ovoid shape of the typical human sperm head makes subtle deviations in shape harder to perceive. Large interlaboratory differences exist in morphology criteria, and in general, there is little agreement on what constitutes a "normal" sperm (Freund, 1966; Fredricsson, 1979). However, studies of MacLeod (1974), David et al. (1975), Eliasson (1971), Wyrobek et al., (1982) and others have shown that quantitative approaches to the visual assessment of morphology can be used with considerable success. Human sperm morphology has been used to study the effects of approximately 40 different chemical exposures (for a review see Wyrobek et al., 1983a). In addition, there has been some recent success in the mouse in removing observer subjectivity with automated image analysis techniques (Young et al., 1982; Moore et al., 1982).

Though the mechanisms await clarification, there is increasing evidence that induced changes in sperm shape may be related to induced genetic damage and impaired fertility. In the mouse, which has been studied in detail, it has been shown that an agent's ability to induce sperm-shape abnormalities is related to its ability to act as a germ-cell mutagen, as judged by dominant lethal, heritable translocation, or specific locus tests (Wyrobek et al., 1983b). In the human literature the relationship between semen quality and adverse reproductive outcome is ambiguous (see Section V).

Certain categories of sperm-shape abnormalities may be related to reduced fertility in man (e.g., Nistal et al., 1978). However, it is not yet clear whether all the categories that are classed abnormal or non-oval in shape consist of sperm that are also abnormal in their fertilizing capacity. In the mouse we know that the shape of the normal sperm, as well as the percent and types of abnormalities seen are remarkably genotype-specific (Wyrobek et al., 1976; Wyrobek, 1979; Krzanowska, 1981). Likewise, individual men seem to produce remarkably consistent proportions and patterns of sperm shapes, even when sampled over a period of more than a year (MacLeod, 1965a; A. Wyrobek, L. Gordon, G. Watchmaker, and D. Moore II, in preparation). Human sperm morphology can be visually scored reliably, and has a greater statistical power as a test of induced chemical effects on sperm production than does sperm count (Wyrobek et al., 1982).

D. Double F-Bodies

The double F-body test is based on scoring the frequency of sperm with 2 fluorescent spots in seminal smears stained with quinacrine dye. Studies in somatic cells with single spots suggest that these spots represent Y chromosomes. Though it has been suggested that sperm with 2 spots contain two Y chromosomes due to meiotic nondisjunction (Kapp, 1979), there remain major uncertainties in this interpretation. Research on this test has been hampered because unlike the other sperm tests (counts, motility, and morphology), the double F-body test has no direct counterpart in common laboratory or domestic animals. The Y-chromosomal fluorescence seems to be unique to man and the higher apes (Seuànez, 1980). It is interesting to note that some field voles, Microtus, have sex chromosomes that are uniquely heterochromatic and can be readily identified in haploid spermatids. This system may provide a promising animal model of chemically induced nondisjunction in male germ cells (Tates, 1979).

The double F-body test is very new and only a few populations of exposed men have been analyzed with it (Kapp et al., 1979; Wyrobek et al., 1981). An evaluation of the scoring criteria and the statistical characteristics of this test are described elsewhere (Wyrobek and Watchmaker, in preparation).

III.   HUMAN MONITORING

The above methods have been applied to assess spermatogenic function in at least 89 different groups of chemically exposed men (Wyrobek et al., 1983a). Tables 1 to 4 categorize these agents into occupational and environmental chemicals (Table 1), experimental and therapeutic drugs (Tables 2 and 3), and recreational drugs (Table 4). Details of the studies surveyed to generate these tables and the decision criteria used to classify each agent as one (a) with adverse effects, (b) suggestive of adverse effects, and (c) with no apparent adverse effects are published elsewhere (Wyrobek et al., 1983a). Several agents (not listed in these tables) have been reported to improve sperm quality in some cases.

IV. GUIDELINES FOR PLANNING A NEW HUMAN SPERM STUDY

Human sperm tests are practical methods for directly assessing spermatogenic damage in the exposed human male. Since sperm tests have been used in a large variety of study designs, and since studies with human subjects must be tailored to given circumstances, it is not appropriate to consider a standardized protocol. The following points should be considered when planning a new study of spermatogenic function in exposed men.

A.   Identifying Populations at Risk

Groups of men may be considered at risk if they are exposed to an agent or close analog of an agent known to be a testicular toxin in any mammal. Men exposed to agents known to be reproductive toxins, mutagens, or carcinogens in male animals should also be considered at risk for spermatogenic damage. Sometimes groups may be identified more directly by the effects of exposure. For example, the study of men exposed to DBCP was initiated by workers' complaints of unintentional childlessness (Whorton et al., 1977). Case reports can be useful

Table 3
Effects of Experimental and Therapeutic Drugs
on Human Sperm: Agents Showing Suggestive
or No Adverse Effects[a]

Agents Suggestive of Adverse Effects	Agents with no Apparent Adverse Effects
Centrochroman	Bromocriptine
Cimetidine	Lysine
Colchicine	Methyltestosterone
Diethylstilbestrol	Niridazole
Methadone	Norethindrone + testosterone
Metronidazole	
Nitrofurantoin	Orinthine
Norethandrolone + testosterone	Tryptophan
	WIN 59,491
Trimeprimine	

[a] For further information see the footnote of Table 1.

Table 4

Effects of Recreational Drug Use on Human Sperm[a]

Agents with Adverse Effects	Agents Suggestive of Adverse Effects	Agents with No Apparent Adverse Effects
Alcoholic beverages (chronic alcoholism)	Tobacco smoke	None reported
Marijuana		

[a] For further information see the footnote of Table 1.

Table 1
Effects of Occupational and Environmental Chemicals on Human Sperm[a]

Agents with Adverse Effects	Agents suggestive of Adverse Effects	Agents with No Apparent Adverse Effects
Carbon disulfide	Carbaryl	Anesthetic gases
Dibromochloropropane	Kepone	Epichlorohydrin
Dibromochloropropane + ethylene dibromide		Glycerine production compounds
Lead		Polybrominated biphenyls
Toluene diamine + dinitrotoluene		

[a] Table entries are based on studies of sperm counts, motility, morphology, and double F-bodies. The assignment of individual agents to columns is based on the data provided in the papers reviewed by the Human Sperm Reviewing Committee of the U.S. Environmental Protection Agency (EPA) Gene-Tox Program (Wyrobek et al., 1983a). These entries are generally based on few studies and may be expected to change as more data become available.

Table 2
Effects of Experimental and Therapeutic Drugs on Human Sperm:
Agents or Combinations of Agents with Adverse Effects[a]

Acridinyl anisidide
Adriamycin
Aspartic acid
Clorambucil
Clorambucil +
  mechlorethamine +
  azathioprine
Clomiphene citrate
Cyclophosphamide
Cyclophosphamide +
  colchicine
Cyclophosphamide +
  prednisone
Cyclophosphamide +
  prednisone +
  azathioprine
CVP (cyclophosphamide +
  vincristine +
  prednisone)
CVPP (cyclophosphamide +
  vincristine +
  procarbazine +
  prednisone)
Cyproterone acetate
Danazol +
  methyl testosterone
Danazol +
  testosterone enanthate
Enovid
Gossypol
Luteinizing hormone
  releasing factor agonist
Medroxyprogesterone acetate
Medroxyprogesterone acetate +
  testosterone enanthate
Medroxyprogesterone acetate +
  testosterone propionate
Megestrol acetate +
  testosterone
Metanedienone
Methotrexate
MOPP (Mechlorethamine +
  vincristine +
  procarbazine +
  prednisone)
MVPP (Mechlorethamine +
  vinblastine +
  prednisolone +
  procarbazine)
Norethandrolone
Norethindrone
Norethindrone +
  norethandrolone +
  testosterone
Norgestrel +
  testosterone enanthate
Norgestrienone +
  testosterone
Prednisolone
Propafenon
R-2323 + testosterone
Sulphasalazine
Testosterone
Testosterone cyclopentyl-
  propionate
Testosterone enanthate
Testosterone propionate
VACAM (Vincristine +
  adriamycin +
  cyclophosphamide +
  actinomycin D +
  medroxyprogesterone acetate)
WIN 13099
WIN 13099 +
  diethylstilbestrol
WIN 17416
WIN 18446

[a] For further information see the footnote of Table 1.

indicators of potential problem exposures, provided caution is used not to overinterpret the data.

B. Study Design

A peer-reviewed protocol is essential. It should cover the study design calculations of statistical power, confidentiality of medical records, informed consent of subjects, instructions for semen collection procedures, instructions for sample handling and transport, and the method of reporting results and interpretations back to the subject. For patients undergoing health care, there should be a written assurance that participation in a sperm study is voluntary and that involvement will in no way affect medical care. Provisions should be made for the follow-up of abnormal test results.

Information on the number of men exposed to the agent as well as any available estimates of dose are essential to planning an effective study. Since between-male variability in semen characteristics is high, large numbers of subjects are required to establish differences between control and exposed groups in cross-sectional studies in which each individual is sampled only once (Wyrobek, 1981; Wyrobek et al., 1982). Cross-sectional studies are therefore typically large, e.g., approximately 170 men in the study of smoking effects (Viczian, 1969), and 200 in the study of lead effects (Lancranjan et al., 1975).

Variation of sperm morphology within an individual male is considerably less than variation among individuals (MacLeod, 1965a, 1974; Sherins et al., 1977). Therefore, prospective longitudinal study designs may be more appropriate when fewer men are available for sampling. In these designs, repeated semen samples in the same man before, during, and after exposure are compared to assess chemically induced sperm defects. Longitudinal studies have been successfully conducted on men exposed to x rays and on men receiving drugs for medical reasons (MacLeod, 1965a; Heller et al., 1966; Rowley et al., 1974).

For evaluation of the side effects of therapeutic agents, various study designs have been used: cross-sectional, case reports, or case-control studies. Unlike the occupational/environmental studies, exposure dose is usually obtainable from medical records. In situations in which a drug with potential testicular toxicity is necessary for treatment of a specific condition, a prospective longitudinal study can be done. Studies of drug effects are generally easier to arrange than occupational studies because of the physician's established relationship with the patient.

C. Gaining Access

Once a population at risk is identified, the investigators need to obtain access to the population. This can be the most time-consuming aspect of the study and may require cooperation from hospital administrators, plant management, labor unions, and various governmental agencies. When industrial chemicals are involved, access may become a major issue since industry is concerned about outside interference in matters that may affect their productivity and economics. Other common problems are: difficulties in obtaining subject cooperation, lack of an available physician to assist in recruitment for drug-related studies, a frequent high percentage of vasectomies among the test population (as high as 25% in some areas of the United States), and refusal for personal reasons.

Numerous methods for recruiting volunteers are possible (e.g., an announcement at general meetings, the circulation of a form letter, physician-patient interaction, etc.). Regardless of the method of introduction, an effective procedure includes a session between a physician and prospective donor on matters of information, clarification, and motivation.

D. Identifying a Control Group

Recruiting concurrent controls is usually more difficult than recruiting exposed men. For occupational studies, control groups typically include (a) new hires, (b) administrative persons, (c) workers from another area of the plant, (d) workers in another plant, and (e) historical controls (i.e., data obtained from men who served as controls in previous studies). Usually each of these groups has some disadvantage due to mismatching in age, socioeconomic factors, numbers of men involved, other chemical exposures, etc. A conservative approach is to use two control groups: the best available concurrent control group and a historical group.

For drug studies, unexposed men may be used as concurrent controls. For longitudinal prospective studies, semen samples collected before exposure may be used as controls.

E. Assigning Men to Dosage Groups

In any study the assessment of dose is vital. Dose-dependent response should be an important criterion in identifying a testicular toxin, and every effort should be made to group men according to dose. Since chemical dosimetry in the workplace is often inadequate, indirect methods of dose grouping need to be considered. For example, Lancranjan et al. (1975) grouped men according to the lead concentrations in urine and blood. Men can also be grouped by job description or number of hours on the job, which may signify exposure level.

Assignment of men to exposure groups is considerably easier in drug studies. With chemotherapy, high doses of chemicals are usually used and recorded in detail. Although such studies may be useful for validating the sperm tests, interpretation is often complicated because combinations of agents were administered and because the disease itself may affect sperm production. Instances where only one agent is used and numerous men are exposed are of particular interest. A prospective sampling of sperm before, during, and after treatment could provide details on dose response, individual susceptibilities to induced damage, and recovery.

F. Collecting Questionnaire Data

1. Types of information needed. Detailed questionnaires are needed to obtain occupational information (dates of employment, types of jobs held, types of agents used, etc.), with special emphasis

on the year prior to semen collection. Detailed medical histories are needed because ill health, recent febrile illness, and associated drug consumption, as well as anatomic abnormalities (as determined by the examining physician) may affect seminal quality. Personal data usually include age, smoking and drinking habits, number and birth dates of children, as well as the occurrence of miscarriages, abortions, and birth defects.

2. <u>Confounding variables</u>. The following is a brief description of factors other than chemical exposure that have been reported to affect human sperm production. They should be considered in the design of the questionnaire and by the physician if a physical examination is done.

   a. <u>Medical factors</u>. Medical conditions reported to affect sperm production in man include: parotitic orchitis (Bartak <u>et al</u>., 1968; (Bartak, 1973), leprosy (Ibrahiem <u>et al</u>., 1979), urogenital tuberculosis (Jimenez-Cruz <u>et al</u>., 1979), renal transplantation (Lingardh <u>et al</u>., 1974), juvenile and adult diabetes (Bartak <u>et al</u>., 1975; Bartak, 1979), prostatitis (Yunda <u>et al</u>., 1978), genito-urinary infection (Caldamone and Cockett, 1978), varicocele (Czyglik <u>et al</u>., 1973; Rodriguez-Rigau <u>et al</u>., 1978; Gall <u>et al</u>., 1978; Greenberg, 1977; MacLeod, 1965b, 1969), severe allergic response (MacLeod, 1965a), febrile and viral diseases (MacLeod, 1965a), and cancer (Wyrobek <u>et al</u>., 1980; Chapman <u>et al</u>., 1981). Men from infertile marriages tend to have lower semen quality than do fertile men (MacLeod, 1951).

   b. <u>Age</u>. Although the effects of age on sperm quality are generally considered to be insignificant, some suggestion of small age-related effects is seen in males of sterile marriage (Atanasov and Tsankov, 1974), or in men with andrological diseases (Schirren, <u>et al</u>., 1975).

c. <u>Radiation</u>. The spermatotoxic and mutagenic effects of radiation on animal testes are well known. Similar patterns of spermatogenic cell killing and subsequent reductions in sperm numbers have been reported for men receiving testicular radiation (Rowley <u>et al</u>., 1974; Greiner and Meyer, 1977).

d. <u>Personal habits</u>. Some relatively common personal habits may affect human sperm production, including tobacco smoking, marijuana use, and heavy alcohol consumption. Exposure to heat, and associated elevations in scrotal temperature, should also be considered (MacLeod and Hotchkiss, 1941; Procope, 1965; Hendry, 1976). Sas and Szöllösi (1979) showed that professional drivers have poorer sperm quality than controls. There has been a suggestion that major dietetic differences (i.e., vegetarian vs nonvegetarian) may be related to seminal quality (Arora <u>et al</u>., 1961).

G. Collecting and Analyzing Semen Samples

In a cross-sectional study, one semen sample per individual is usually analyzed. Longitudinal studies require several samples from each man over a period of several months. Before collection, each donor is given an instruction form emphasizing the importance of obtaining a complete and fresh sample, and explaining the methods of collection. Various methods have been used to collect samples, including masturbation, coitus interruptus, and collection with a seminal pouch. Since count varies with different portions of the ejaculate, the entire ejaculate is necessary to evaluate sperm count. The preferred method is masturbation into a clean glass or plastic container.

The accuracy of some of the semen analyses is critically dependent on the collection and care of the sample. Controlling for continence time is necessary if sperm count is to be accurately evaluated. Morphology, however, does not appear to be affected by continence time (David, 1981). Motility is the only test of the four requiring a fresh sample. Careful control of the time of analysis after sample collection, and sample temperature between collection and analysis is particularly important for the assessment of motility. This control can be easily accomplished if the samples are collected in the clinic, but presents a major difficulty if they are collected at home and brought to the laboratory for analysis. Since home collection greatly improves the participation rate in most studies, the assessment of motility is often not feasible. Although frozen samples can be used to assess sperm counts, morphology, and double F-bodies, it is recommended that fresh samples be used to do the count and make smears whenever possible. Morphology and double F-bodies can be assessed from air-dried smears, which can be shipped and stored for extended periods of time, adding much flexibility to the use of these sperm tests (see Section II for further descriptions and references).

H. Statistical Evaluation and Interpretation

Distributions of sperm data from control and exposed men can be analyzed by parametric and nonparametric methods, e.g., the t test, Mann-Whitney test, Kolmogorov-Smirnov test (as in the DBCP study of Whorton et al., 1977); or by analyses of the proportion of men with semen pathologies (as in the lead study by Lancranjan et al., 1975). Dose relationships should be assessed whenever possible. The possible effects of age, smoking, illness, medication, drugs, and other factors must be analyzed statistically (see item F above for a list of factors to consider).

V. REPRODUCTIVE IMPLICATIONS OF INDUCED SPERM CHANGES

Although it is generally agreed that major reductions in sperm counts and motility are linked to reduced fertility, it is unclear to what extent small reductions are important. It also remains unclear whether changes in any of the human sperm parameters are quantitatively related to embryonic failure, heritable genetic abnormalities, or birth defects. Human data on these questions are extremely limited. We were unable to identify any agent for which we have data in man for the relationships among all four of the following items: (1)

paternal exposure to the chemical, (2) subsequent sperm changes, (3) subsequent fertility changes, and (4) any adverse reproductive outcome such as miscarriages, birth defects, and heritable effects in offspring fathered after exposure. A survey of Tables 1 to 4 emphasizes the relatively large volume of information available on the relationship between items 1 and 2. Chemically induced sperm changes have been linked to reduced fertility (i.e., linking items 1, 2, and 3) for DBCP, cancer chemotherapeutic agents, and several antifertility drugs.

The studies of Furuhjelm et al. (1962) suggest a link between poor sperm quality (count and morphology) and the frequency of embryonic failures (i.e., between items 2 and 4). In these studies, fathers of 201 spontaneous abortions showed significantly higher sperm abnormalities and lower sperm counts than did fathers of 116 normal pregnancies. No data on paternal exposure to any chemical was given. Although several studies support the link between sperm defects and abnormal reproductive outcome (see Czeizel et al., 1967, for a list of references; Joel, 1966; Jöel and Chayen, 1971; Raboch, 1965, for studies on habitual abortions; Furuhjelm et al., 1960, 1962, for studies of fetal loss and perinatal mortality; and Takala, 1957, for studies of birth defects), some studies found no correlation (e.g., Kneer, 1957; MacLeod and Gold, 1957; Homonnai et al., 1980). Clearly, more human studies are needed to compare exposure of the male parent, induced sperm defects, and reproductive outcomes.

Most of the studies on genetic validation of induced sperm anomalies have been conducted in the mouse. The mouse sperm morphology system has been best studied. There is evidence that sperm shape abnormalities are genetically controlled, that induced abnormalities can be related to mutagen exposure, and that changes in sperm abnormalities can be inherited (for a review, see Wyrobek et al., 1983b). Several human studies also support the suggestion that sperm abnormalities may be inherited (Nistal et al., 1978; Bisson et al., 1979).

VI. ROLE FOR SPERM TESTS IN ANIMALS

The availability of animal and human sperm assays suggests several applications in the assessment of chemically induced spermatotoxicity, antifertility effects, and heritable genetic abnormalities. Animal sperm assays (such as sperm count and morphology) may be used to screen large numbers of agents to establish a ranking that sets priorities for sperm studies in exposed men. This approach would minimize the use of human studies, which generally have complex requirements for epidemiological and statistical input and often require lengthy interactions with union officials, industry representatives, employees, physicians, and patient-donors. However, this approach requires a more detailed understanding of the relationship between the response in man and in mouse. It is interesting that there is a large group of exposures for which we have human but no mouse sperm data as yet (Wyrobek et al., 1983b). Animal sperm studies may also be useful in evaluating the relative effects of the components of a complex mixture that is suspected of affecting human sperm (such as in an occupational exposure).

VII. CONCLUSIONS

The major advantages of sperm assays in hazard evaluation are that the cells examined are readily available in both animals and men, and that by studying sperm one is looking at the cell that carries the paternal genome in the form that will be ultimately involved in fertilization. The following are some of the specific advantages of these tests:

1) Sperm are examined after exposure of a whole mammal. This helps ensure that artifacts (false positives and false negatives) due to problems of tissue penetration, metabolism, pharmacokinetics, and dosage, encountered in cultured-cell or nonmammalian systems, are minimized.

2) The changes in sperm parameters probably arise from interference by the test substance with the differentiation of the sperm cell. Thus, these

changes are intrinsically relevant to safety evaluation and assessment of potential effects of the agent on male fertility.

3) The laboratory methods are generally rapid, inexpensive, and quantitative.

4) Sperm tests have clear advantages over other approaches for assessing induced changes in testicular function in man. Testicular biopsies are impractical, traumatic, invasive, and may themselves affect testicular function. Epidemiological surveys of reproductive function using questionnaires exclusively, require large sample sizes and are generally expensive. Analyses of blood levels of gonadotrophins are expensive and generally insensitive to small changes in spermatogenic function. Compared with these methods, sperm tests are noninvasive, less expensive, require small sample sizes for effective analyses, and are sensitive to small changes.

The major disadvantages of sperm assays are the following:

1) The relationship between induced changes in the various sperm parameters and heritable consequences is not yet clearly understood.

2) Limited sperm sampling times and dosage regimens may reduce the sensitivity of the test (e.g., agents that only exert transient effects may be missed by using single sampling times).

3) Other factors such as ischemia, infection, and starvation may produce spurious false positive responses.

VIII. ACKNOWLEDGMENTS

We thank L. Dobson for suggestions in the preparation of this manuscript as well as J.C. Cherniak and A. Riggs for editing, formatting, and typing. We also thank the members of the Gene-Tox Reviewing Committee for Human and Animal sperm tests, whose final reports formed the basis for parts of this manuscript.

IX. REFERENCES

Amelar RD (1966). The semen analysis. In "Infertility in Men: Diagnosis and Treatment," Philadelphia: F.A. Davis Co., pp. 30-53.

Arora RB, Saxena KN, Choudhury MR, Choudhury RR (1961). Sperm studies on Indian men. Fertil Steril 12:365-366.

Ash P (1980). The influence of radiation on fertility in man. Br J Radiol 53:271.

Atanasov A, Tsankov TS (1974). The influence of the age of men from sterile families on several clinical parameters of the ejaculate. Akush Ginekol (Sofiia) 13:199-203.

Babich H, Davis DL, Stotzky G (1981). Dibromochloropropane (DBCP): A review. The Science of Total Environment 17:207-221.

Bartak V (1973). Sperm count, morphology, and motility after unilateral mumps orchitis. J Reprod Fert 32:491-494

Bartak V (1979). Sperm quality in adult diabetic men. Intern J Fertil 24:226-232.

Bartak V, Skalova E, Nevarilova A (1968). Spermiogram changes in adults and youngsters after parotitic orchitis. Intern J Fertil 13:226-232.

Bartak V, Josifko M, Horackova M (1975). Juvenile diabetes and human sperm quality. Intern J Fertil 20:30-32.

Bisson JP, Leonard C, David G (1979). Familial character of some morphological abnormalities of spermatozoa. Arch Anat Cytol Pathol 27:230-233.

Caldamone AA, Cockett ATK (1978). Infertility and genitourinary infection. Urology 12: 304-312.

Chapman RM, Sutcliffe SB, Malpas JS (1981). Male gonadal dysfunction in Hodgkin's Disease. JAMA 245:1323-1328.

Czeizel E, Hancsok M, Viczian M (1967). Examination of the semen of husbands of habitually aborting women. Orvosi Hetilap 108:1591-1595.

Czyglik F, David G, Bisson JP, Jouannet P, Gernigon C (1973). Teratospermia in varicocele. Nouv Presse Med 2:1127-1130.

David G (1981). Factors affecting the variability of semen characteristics. In Spira A, Jouannet P (eds): "Human Fertility Factors (With Emphasis on the Male)." INSERM 103:57-68.

David G, Bisson JP, Czyglik F, Jouannet P, Gernigon C (1975). Anomalies morphologiques du spermatozoïde humain 1) Propositions pour un système de classification. J Gyn Obst Biol Repr 4:17-36.

Eliasson R (1971). Standards for investigation of human semen. Andrologie 3:49-64.

Fredricsson B (1979). Morphologic evaluation of spermatozoa in different laboratories. Andrologia 11:57-61.

Freund M (1966). Standards for the rating of human sperm morphology. A cooperative study. Intern J Fertil 11:97-118.

Furuhjelm J, Jonson B, Lagergren CG, Lindgren L (1960). The quality of the human semen in relation to perinatal mortality. Acta Obstet et Gynec Scandinav 39:499-505.

Furuhjelm M, Jonson B, Lagergren C.G. (1962). The quality of human semen in spontaneous abortion. Intern J Fertil 7:17-21.

Gall H, Schnierstein J, Glowania HJ (1978). The effect of varicocele on male fertility with particular consideration of progressive motility. Urologe 17:317-320.

Greenberg SH (1977). Varicocele and male fertility. Fertil Steril 28:699-706.

Greiner R, Meyer A (1977). Reversible and irreversible azoospermia after irradiation of the malignant tumor of the testicle. Strahlentherapie 153:257-262.

Heller CG, Wootton P, Rowley MJ, Lalli MF, Brusca DR (1966). Action of radiation upon human spermatogenesis. In: Proceedings of the Sixth Pan American Congress of Endocrinology, Mexico City. Excerpta Med Intern Cong Ser No 112:408-410.

Hendry WF (1976). Loose pants and cold scrotal douches: effects on spermatogenesis. Sperm Action Prog Reprod Biol 1:259-262.

Homonnai ZT, Paz GF, Weiss JN, David MP (1980). Relation between semen quality and fate of pregnancy: Retrospective study on 534 pregnancies. Intern J Androl 3:574-584.

Ibrahiem AA, Awad HA, Metawi BA, Hamada TAY (1979). Pathologic changes in testis and epididymis of infertile leprotic males. Intern J Lepr 47:44-49.

Jimenez-Cruz JF, Saenz de Cabezon J, Soler-Rosello A, Sole-Balcells F (1979). The spermiogram in urogenital tuberculosis. Andrologia 11:67-70.

Jöel CA, Chayen R (1971). Pathological semen as a factor in abortion and infertility. In "Fertility Disturbances in Men and Women." Karger: Basel, p 496-507.

Jöel CA (1966). New etiologic aspects of habitual abortion and infertility, with special reference to the male factor. Fertil Steril 17:374-380.

Jouannet P, Volochine B, Deguent P, Serres C, David G (1977). Light scattering determination of various characteristic parameters of spermatozoa motility in a serie of human sperm. Andrologia 9:36-49.

Kapp RW (1979). Detection of aneuploidy in human sperm. Environ Health Perspect 31:27-31.

Kapp RW, Picciano DJ, Jacobson CB (1979). Y-chromosomal nondisjunction in dibromochloropropane-exposed workmen. Mutat Res 64:47-51.

Katz DF, Overstreet JW (1981). Sperm motility assessment by videomicrography. Fertil Steril 35:188-193.

Kneer M (1957). Der habituelle abort. Deutsch Med Wochenschr 82:1059-1060.

Krzanowska H (1981). Sperm head abnormalities in relation to the age and strain of mice. J Reprod Fertil 62:385-392.

Lancranjan I, Popescu HI, Gavanescu O, Klepsch I, Serbanescu M (1975). Reproductive ability of workmen occupationally exposed to lead. Arch Environ Health 30:396-401.

Levine RJ, Symons MJ, Balogh SA, Milby TH, Whorton MD (1981). A method for monitoring the fertility of workers 2. Validation of the method among workers exposed to dibromochloropropane. J Occup Med 23:183-188.

Lingardh G, Andersson L, Osterman B (1974). Fertility in men after renal transplantation. Acta Chir Scand 140:494-497.

MacLeod J (1974). Effects of environmental factors and of antispermatogenic compounds on the human testis as reflected in seminal cytology. In: Mancini RE, Martini L (eds): "Male Fertility and Sterility." Proc Serono Symp, Vol. 5. New York, Academic Press, p 123-148.

MacLeod J (1951). Semen quality in one thousand men of known fertility and in eight hundred cases of infertile marriage. Fertil Steril 2:115.

MacLeod J (1965a). Human seminal cytology following the administration of certain antispermatogenic compounds. In Austin CR, Perry JS (eds): "Symposium on Agents Affecting Fertility," Little, Brown and Co., Boston, p 93-123.
MacLeod J (1965b). Seminal cytology in the presence of varicocele. Fertil Steril 16:735-757.
MacLeod J (1969). Further observations on the role of varicocele in human male infertility. Fertil Steril 20:545-563.
MacLeod J, Hotchkiss RS (1941). The effect of hyperpyrexia upon spermatozoa counts in men. Endocrinology 28:780-784.
MacLeod J, Gold RZ (1957). The male factor in fertility and infertility: IX Semen quality in relation to accidents of pregnancy. Fertil Steril 8:36.
Makler A, Zaidise I, Paldi E, Brandes JM (1979). Factors affecting sperm motility. I. In vitro change in motility with time after ejaculation. Fertil Steril 31:147-154.
Moore DH II, Bennett DE, Kranzler D, Wyrobek AJ (1982). Quantitative methods of measuring the sensitivity of the mouse sperm morphology assay. Anal Quant Cytology 4:199-206.
Morgan RW, Kheifets L, Obrinsky DL, Milby TH, Whorton MD (1983). Fetal loss and work in a waste water treatment plant. American J Public Health (in press).
Nistal M, Harruzo A, Sanchez-Corral F (1978). Toratozoospermia absoluta de presentacion familiar. Espermatozoides microcefalos irregulares sin acrosoma. Andrologia 10:234.
Procope BJ (1965). Effect of repeated increase of body temperature on human sperm cells. Intern J Fertil 10:333-339.
Raboch J (1965). Spermiologic findings in repeated spontaneous abortion. Zentralbl Gynaekol 87:194-197.
Rodriguez-Rigau LJ, Smith KD, Steinberger E (1978). Relationship of varicocele to sperm output and fertility of male partners in infertile couples. J Urol 120:691-694.
Rowley MJ, Leach DR, Warner GA, Heller CG (1974). Effect of graded doses of ionizing radiation on the human testis. Radiat Res 59:665-678.
Sas M, Szöllösi J (1979). Impaired spermiogenesis as a common finding among professional drivers. Archives of Andrology 3:57-60.

Schirren C, Laudahn G, Hartmann E, Heinze I, Richter E (1975). The correlation of morphological and biochemical factors in human ejaculate in various andrological diagnoses I. Relationship between ejaculate volume, number, motility, and morphology of the spermatozoa with regard to age. Andrologia 7:117-125.

Schwartz D, Laplanche A, Jouannet P, David G (1979). Within-subject variability of human semen in regard to sperm count, volume, total number of spermatozoa, and length of abstinence. J Reprod Fertil 57:391-395.

Seuànez HN (1980). Chromosomes and spermatozoa of the African great apes. J Reprod Fertil Suppl 28:91-104.

Sherins RJ, Brightwell D, Sternthal PM (1977). Longitudinal analysis of semen of fertile and infertile men. In Troen P, Nankin HR, (eds): "The Testis in Normal and Infertile Men," New York: Raven Press, p 473-488.

Takala ME (1957). Studies on the seminal fluid of fathers of congenitally malformed children (199 sperm analyses). Acta Obst et Gynec Scandinav 36:29-41.

Tates AD (1979). Microtus oeconomus (Rodentia), a useful mammal for studying the induction of sex-chromosome nondisjunction and diploid gametes in male germ cells. Environ Health Perspect 31:151-159.

Viczian M (1969). Ergebnisse von spermauntersuchungen bei zigarettenrauchern. Zschr Haut-Geschl-Krkh 44:183-187.

Whorton D, Krauss RM, Marshall S, Milby TH (1977). Infertility in male pesticide workers. Lancet 2:1259-1261.

Whorton D, Milby TH, Krauss RM, Stubbs HA (1979). Testicular function in DBCP exposed pesticide workers. J Occup Med 21:161-165.

Wyrobek AJ (1979). Changes in mammalian sperm morphology after x-ray and chemical exposures. Genetics 92:S105-S119 supp.

Wyrobek AJ, da Cunha MF, Gordon LA, Watchmaker G, Gledhill B, Mayall B, Gamble J, Meistrich M (1980). Sperm abnormalities in cancer patients. Proc Am Assoc Cancer Res 21:196.

Wyrobek AJ, Gordon LA, Burkhart JG, Francis MW, Kapp RW, Letz G, Malling HV, Topham JC, Whorton D (1983a). An evaluation of human sperm as indicators of chemically induced alterations of spermatogenic function: A report of the U.S. Environmental Protection Agency Gene-Tox Program. Mutat Res 115:73-148.

Wyrobek AJ, Gordon LA, Burkhart JG, Francis MW, Kapp RW, Letz G, Malling HV, Topham JC, Whorton D, (1983b). An evaluation of the mouse sperm morphology test and other sperm tests in nonhuman animals: a report of the U.S. Environmental Protection Agency Gene-Tox program. Mutat Res 115:1-72.

Wyrobek AJ, Gordon LA, Watchmaker G, Moore II DH (1982). Human sperm morphology testing: Description of a reliable method and its statistical power. In: Banbury Report 13: Indicators of Genotoxic Exposure. Cold Spring Harbor Laboratory, p. 527-541.

Wyrobek AJ, Meistrich ML, Furrer R, Bruce WR (1976). Physical characteristics of mouse sperm nuclei. Biophysical J 16:811.

Wyrobek AJ, Watchmaker G, Gordon L, Wong K, Moore D II, Whorton D (1981). Sperm shape abnormalities in carbaryl-exposed employees. Environ Health Perspect 40:255-265.

Young IT, Gledhill BL, Lake S, Wyrobek AJ (1982). Quantitative analysis of radiation-induced changes in sperm morphology. Anal Quant Cytology 4:207-216.

Yunda If, Imshinetskaja LP, Karpenko EI, Tschernyschow WP, Sokolowa MN, Gorpintschenko II (1978). Prostatitis and pathospermia. Dermatol Monatsschr 164:564-567.

Evaluating Male Reproductive Toxicity in Rodents:
A New Animal Model

H. Zenick, K. Blackburn, E. Hope, D. Oudiz, H. Goeden
Department of Environmental Health,
University of Cincinnati
Cincinnati, OH 45267

Concern about the susceptibility of the male reproductive system to environmental agents is relatively recent. Much of this interest was generated by studies of workers exposed to dibromochloropropane (e.g. Whorton, et al., 1979). In this instance reduced fertility, as a result of oligozoospermia, was present without any other clinical signs of toxicity. These findings suggested that for certain agents the reproductive system could be the first or most sensitive target organ. Besides the issue of fertility, there is the prospect of paternal exposure contributing to preimplantation or fetal loss, birth defects, childhood cancer, or neurobehavioral deficits (see papers by Adams and Peters this volume).

The task of assessing reproductive risk in human populations is complicated by several factors. First, almost all information on sperm integrity (i.e. sperm count, viability, morphology) has been obtained from suspected infertile or subfertile men. In studies of presumably healthy, fertile men, limited data are available relating sperm integrity to sperm competence (i.e. fertilizing ability, successful pregnancies). If such information is lacking for fertile, normal populations, it follows that the impact of environmental agents on sperm integrity or competence is even more obscure.

Even assuming a defined relationship between sperm integrity and sperm competence, the study of environmentally-exposed human populations is fraught with problems. Gaining access to populations may be difficult.

© 1984 Alan R. Liss, Inc.

In addition large numbers of individuals are required to detect exposure-related alterations in reproductive outcomes (e.g., early fetal loss). Additional difficulties relate to the individual's simultaneous exposure to multiple agents and uncertainty in defining actual exposure levels. Finally, the opportunity to conduct prospective, long-term studies relating exposure to semen status and concurrent reproductive activities (outcomes) is rarely afforded. Yet this is the ideal design in which to study these relationships.

As a result of the restrictions encountered in the conduct of human reproductive studies, data from animal studies may have to be used to identify potential chemical hazards and to suggest exposure thresholds for adverse effects. Such data would ideally aid in the design of subsequent human studies as well as the delineation of mechanisms underlying the human response. However, most of the existing animal studies have evaluated only reproductive outcomes and not sperm integrity. Yet, it is the latter that provides endpoints most readily assessed in the human population.

Some of the parameters that can be evaluated in animal studies of the male reproductive system are listed in Table I. Libido and potency have been traditionally evaluated in an indirect manner. Males and females are placed together overnight, the presence or absence of a copulatory plug is determined the following morning. Occasionally, a vaginal wash is performed to establish the presence of sperm. This testing regimen, which evaluates only the presence of a plug (or sperm), cannot differentiate between reproductive organ dysfunction (i.e., failure to produce sperm) and CNS impairment (changes in libido). The presence of a copulatory plug (or sperm in the vaginal wash) does not insure that the copulatory process is unaffected, but only that during the night the male was able to achieve at least one ejaculation. The sole means of evaluating the functional significance of CNS dysfunction on reprodutive performance is to monitor the copulatory sequence.

The importance of such assessments in the course of reproductive evaluations should not be understated. Numerous agents (e.g., solvents) produce CNS disturbances that have the potential to interfere with copulatory behaviors and subsequent successful impregnation. This latter point is supported by research showing that the ejaculatory process contributes to the number of sperm ejaculated and to sperm transport in the female (Chester and Zucker, 1970). Data from this laboratory on trichloroethylene (TCE) and carbon disulfide (CS_2) have shown that alterations in copulatory behaviors can be the earliest indicators of toxicity among the reproductive parameters examined. These effects are discussed later in this paper. Interestingly, both of these agents have been associated with disturbances in sexual dynamics in occupationally exposed workers.

The second category of Table 1 lists measures related to the structural and functional integrity of the reproductive organs themselves. The predominant measure employed has been histopathological evaluation of the testes. This endpoint is important in providing insight into site(s) of insult. However, such techniques are not applicable to the study of human populations. Furthermore, the lack of testicular lesions does not preclude reproductive impairment. For example, secretions of the accessory glands nourish and maintain the spermatozoa. Thus, an impairment of the accessory organs could affect sperm integrity.

The number of studies in which sperm evaluations have been conducted is limited. Such evaluations are usually performed on sperm recovered from the cauda epididymis at time of sacrifice. Endpoints assessed include sperm count and morphology, and occasionally, sperm motility. The predominant sperm test in non-human mammals has been the mouse sperm morphology test used as an indicator of germ cell mutation. However, its utility when applied to environmental agents and/or tested in other species remains unconfirmed. The marked absence of evaluation of other parameters of spermatogenic dysfunction is best documented in a recent review by the EPA Gene-Tox Program (1983).

The final category of fertility and fetal outcome noted in Table 1 has been primarily evaluated using one of

TABLE I

PARAMETERS THAT CAN BE EVALUATED IN MALE REPRODUCTIVE STUDIES

I. LIBIDO AND POTENCY
   1. Endocrine primary lesions
   2. CNS (non-endocrine) lesions

II. EVALUATIONS OF THE REPRODUCTIVE SYSTEM
   a. Measures of sperm characteristics (concentration, viability, motility, morphometry, fertilization potential)
   b. Biochemical markers (enzymes, proteins, unscheduled DNA synthesis)
   c. Histopathological assessment
   d. Status of accessory organs and hormone analysis

III. FERTILITY AND FETAL OUTCOMES
   a. Infertility
   b. Pre-, post-implantation loss
   c. Fetal viability and survival
   d. Postnatal status
      1. Survival 2. Growth
      3. Functional deficits

three basic strategies. The first is female-only exposures, either through organogenesis with pre-term evaluation of fetuses (teratology) or through both gestation and lactation with some postnatal evaluation of the offspring. The second is simultaneous exposure of males and females through 2-3 generations with emphasis on the reproductive performance of the F1 and/or F2 generations. In this instance paternal versus maternal influences cannot be distinguished. The final approach entails tests for germ cell mutation, exemplified in mammalian systems by the dominant lethal test. These tests are designed primarily to screen agents for their ability to cause embryonic death via germ cell mutation.

In summary, there are not any existing standardized testing strategies to evaluate the function of the male reproductive system, in terms of either fertility or fetal outcome, for agents whose primary mode of action is not germ cell mutation. Moreover, since most species produce a superabundance of sperm, fertility and fecundity are not particulary sensitive criteria for monitoring the reproductive process (Amann, 1982).

We have attempted to develop an animal model that will provide information on the overall function of the male reproductive system (Table 2). As outlined earlier, data of this nature are not provided by other testing strategies. In addition, we have emphasized endpoints which can be applied to studies of human reproductive function. The approach entails evaluating ejaculated semen samples recovered from the reproductive tract of a female rat at specified times post-copulation. This approach has a number of strengths:

1) The recovered ejaculate can be evaluated for semen parameters that may also be assessed in the human population.

2) For a given animal, a semen evaluation may be conducted prior to exposure. Repeated assessments may then occur during exposure as well as post-exposure (recovery). To this end, each animal may serve as his own control, enhancing the probability of detecting treatment related alterations. This advantage can be best appreciated when one considers the inter-individual variability associated with semen evaluations. Moreover, a clearer

TABLE II. PARAMETERS EXAMINED IN REPEATED MEASURES MODEL

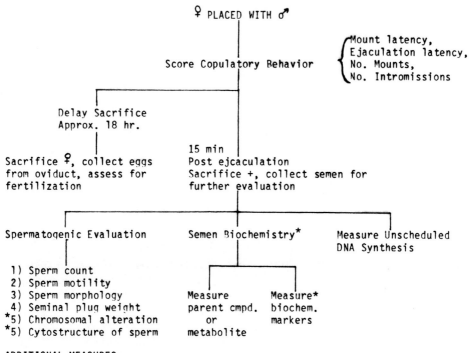

ADDITIONAL MEASURES:

1) Endocrine evaluation on ♂
2) Fertility assessment
3) Teratogenic or postnatal assessments

*Under development

picture of the degree of recovery for a given animal can be ascertained.

3) Serial matings can be conducted concurrent with semen evaluations so that changes in the functional integrity of the sperm and reproductive competence can be correlated.

4) Traditional methods of semen recovery in rodents (e.g., electro-ejaculation) are stressful and may not be repeatedly applied. Our strategy of examining ejaculates recovered from the reproductive tract of a receptive female avoids this problem. The design also allows us to obtain data on copulatory behavior and sperm integrity while preserving the most natural environment in which the sperm may reside until evaluation.

This model has been applied to the study of a number of compounds including TCE, 2,4,6-trichlorophenol, CS_2, and 2-ethoxyethanol (2-EE). We have also used this approach to monitor unscheduled DNA synthesis (UDS) in the ejaculate following exposure to methylmethane sulfonate (MMS). In the remainder of this paper we will describe some of the aspects of the model and present normative and treatment-related data validating its utility. Data are also presented on our attempts to better quantitate certain semen parameters (e.g., motility) by using a computer-analyzed, videomicrographic scoring system. This system is also being used to evaluate human semen samples (Katz and Overstreet, 1981).

The timetable routinely employed in conjunction with this model is indicated in Table 3. Male and female rats (Long Evans hooded, Charles River) are introduced into the laboratory at 70 days of age. Females are then ovariectomized and allowed a one-to-two week recovery period. Receptivity is induced by an injection of estradiol 48 hrs prior to mating (0.1 mg/ml), followed by an injection of progesterone (0.1 mg/ml) 4 hrs before mating. Initial investigations have found no differences in semen characteristics in samples recovered from intact, estrous females as contrasted to ovariectomized, hormonally-primed animals (Fig. 1).

TABLE III. TIMETABLE

Age

70 Days	♂ Introduced into laboratory
	Mated weekly with ♀ to obtain mating proficiency
100 days	Obtain baseline on copulatory behaviors and semen parameters
107 days	Begin exposure
114 days	Evaluate Wk 1
135 days	Evaluate Wk 4
156 days	Evaluate Wk 7
177 days	Evaluate Wk 10
	Continue Exposure Histology Monitor Recovery
	Fertility Assessment

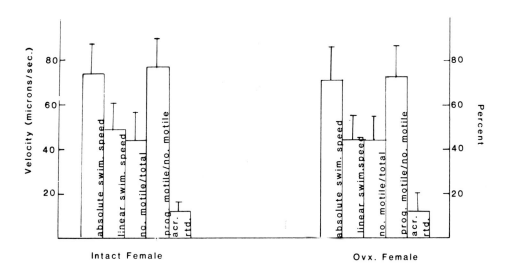

Fig. 1.

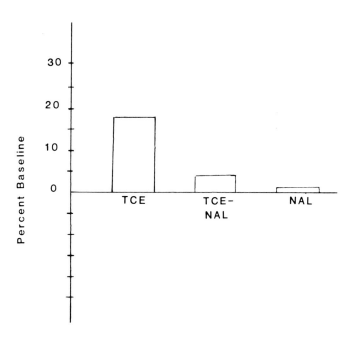

Fig. 2.

During the initial 30-day period males are mated several times in order to gain proficiency in copulatory behavior. At 100 days of age, a baseline evaluation of sperm parameters and copulatory behaviors is conducted. Males with extremely low sperm counts ($\leq$ 20 million/ml) or protracted ejaculation latencies (> 20 minutes) are eliminated from study. The remaining males are ranked on these two variables and assigned to control and treatment groups in such a manner as to produce comparable distributions across groups prior to initiating treatment.

The initial evaluation of semen and mating behavior is performed at 100 days of age, since at this time sperm production has reached its maximum adult level (Saksena, et al., 1979). Studies that initiate exposure at 60-70 days of age risk introducing confounding factors attributable to the differing sensitivity of immature animals to insult. Insults incurred at this age may be quite different from those seen in the mature adult.

Because of the periodicity of the spermatogenic cycle, an acute (5 days) treatment can be used to pinpoint effects on specific spermatogenic stages. Semen evaluations are conducted at one, four, seven and ten weeks post-exposure. Effects seen at specific time points would be indicative of damage to the spermatozoa, spermatid, spermatocyte, or spermatogonia stage, respectively. Subchronic exposures are usually for 70-80 days, corresponding to one full cycle of spermatogenesis. Several options may be pursued at the end of this period, including sacrifice accompanied by histological evaluation, traditional fertility testing and/or monitoring of recovery. Males are mated weekly, even when semen evaluations are not conducted, in order to insure maintenance of a constant abstinence period.

As can be seen in Table 2, the first component evaluated is copulatory behavior. The male is placed into a Plexiglas® observation chamber and given a 15 minute adaptation period. All observations are conducted under red light and during the dark phase of the light-dark cycle. The female is subsequently introduced and four behaviors scored: 1) mount latency (the interval between the introduction of the female and first mount), 2) the number of mounts, 3) the number of intromissions, and 4) ejaculation latency (the interval between the first mount and ejaculation). The occurrence of ejaculation in a rat is easily confirmed by presence of a seminal plug in the female tract at sacrifice. As noted earlier, two of the compounds we have studied to-date, TCE and CS_2, alter copulatory behavior. A single TCE treatment (1000 mg/kg, p.o.) produces a protracted ejaculation latency which may be consistent with its purported narcotic properties. To this extent treatment with the narcotic antagonist, naltrexone (10 mg/kg, i.p.) reverses this behavior (Fig. 2). Moreover, daily administration of TCE (1000 mg/kg, p.o.) produces comparable narcotic effects in the initial weeks of treatment. However, the effect on copulatory behavior is absent by five weeks of exposure. This phenomenon may be comparable to tolerance observed with continuous narcotic injections.

Daily CS_2 exposure (600 ppm, inhalation) produced an opposite effect, namely a significant decrease in ejaculation latencies seen by the fourth week of exposure (Fig. 3). These same males exhibited declines in ejaculated sperm counts in subsequent weeks. Interestingly, cauda epididymal sperm counts were not altered in these animals. It is possible that CS_2-induced alterations in copulatory behavior (e.g., premature ejaculation?) may have affected the number of sperm ejaculated (Chester and Zucker, 1970). Alternatively, CS_2 exposure may interfere with contractility of the vas deferens, reducing the number of sperm ejaculated. The important point is that conventional approaches which examine only cauda epididymal reserves would not have detected this effect.

Subsequent to copulation, the female remains undisturbed for 15 minutes and then is sacrificed (CO_2 asphixiation). If intact estrous females are employed, sacrifice can be delayed for eighteen hours, and the eggs recovered from the ampulla and examined for signs of fertilization. To recover the semen sample, an abdominal incision is made, the reproductive tract exposed and the uterine contents withdrawn into a syringe (37°C). The tract is then excised, the seminal plug removed, rinsed, weighed and the remainder of the seminal fluid flushed from the tract. Seminal plug weight is an easily obtained measure and may serve as a marker of hormone status since it is a product of androgen-dependent glands. An evaluation of 141 Long Evans hooded, male rats at approximately 100 days of age shows a mean seminal plug weight of 0.103 ± 0.021 gm (range 0.032 gm to 0.188 gm). As animals mature, seminal plug weights generally increase.

An aliquot of the syringe contents is diluted with culture medium (37°C) to provide a sample dilute enough to be scored for motility. The sample is then placed on a slide and several frames videotaped. The tape can then be played back for subsequent evaluation of motility.

A 15 ul aliquot of the diluted sample is smeared on two microscope slides, which are air-dried and stained with a combination of eosin Y, fast green and naphthol yellow (Bryan, 1970). These slides are used for morphology evaluations. In our experience, the LEH rat shows less than 3% abnormal sperm, in both ejaculated and epididymal samples. This combination of stains can also be used to visualize the acrosome. This is an important consideration, as an intact acrosome is necessary for subsequent capacitation and fertilization.

Total sperm number is obtained by rinsing all microscope slides, syringe contents, and tract washings with distilled water and diluting to a standard volume (50 ml). Aliquots of this dilution are loaded in both chambers of a Neubauer hemocytometer and two cell counts are performed. Data on the distribution of sperm count in control animals obtained from the ejaculate and from the cauda epididymis are presented in Fig. 4 and Fig. 5, respectively.

The videotape is evaluated for percent motile sperm. In addition, fifty sperm in various segments of the tape are scored for swimming pattern and distance travelled (absolute distance and linear distance). The distance variables are measured using a digitizing cursor to track the sperm. These data are fed into a computer along with elapsed time to generate swimming speeds (microns/sec). A frequency distribution of swimming speeds and patterns is then generated for each male along with average swimming speed. Figures 6a, 6b and 6c reflect sample distribution curves in the rat for linear and absolute swimming speeds and the absolute/linear ratios respectively. The latter

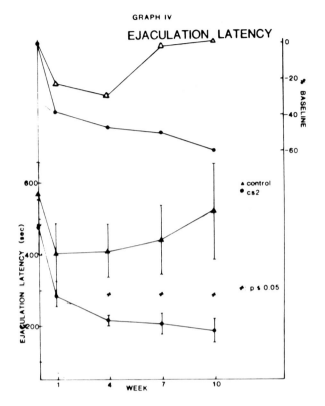

Fig. 3.

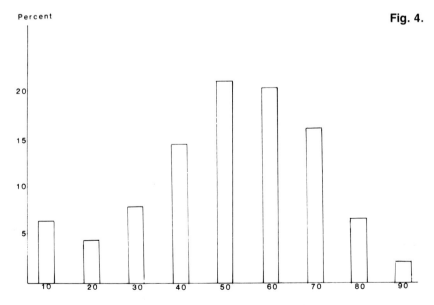

Fig. 4.

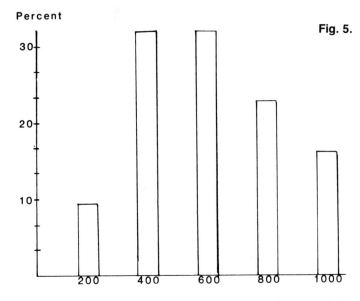

Fig. 5.

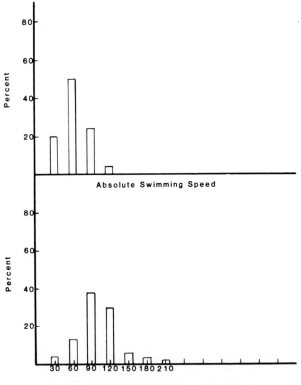

Fig. 6a.

Fig. 6b.

Male Reproductive Toxicity 123

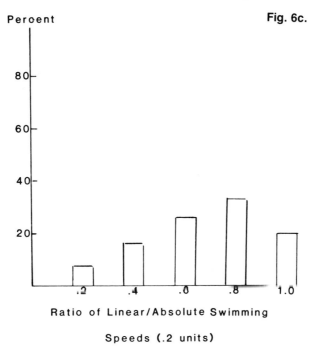

Fig. 6c.

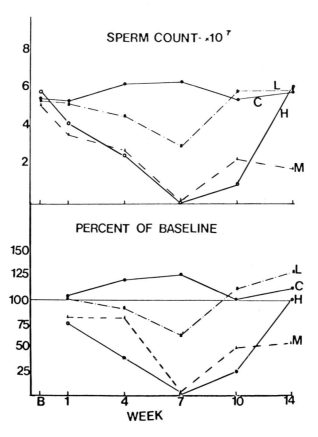

Fig. 7.

also takes into account swimming pattern in the expression of the final ratio. We feel that this approach maximizes the data generated from a semen sample. The distribution of each animal may then serve as a "sperm print" to contrast against the normative data base. Theoretically, specific exposures (xenobiotics, diseases, pathologies, etc.) could be identified by such sperm profiles. However, this theory remains speculative until sufficient data are generated from future studies.

Data from a recent investigation of 2-ethoxyethanol (2-EE) can be used to illustrate the application of our model in evaluating various sperm parameters. In that study, a baseline evaluation was conducted; male rats then received either 0, 936, 1872, or 2808 mg/kg (p.o.) of 2-EE for five consecutive days. The males were mated weekly for the next 14 weeks. Semen evaluations were conducted on Weeks 1,4,7,10, and 14. The latter time point was included because males had begun to show recovery by the tenth week post-exposure.

Data analyses indicated that 2-EE produced a rapid decline in sperm counts in the two highest groups, with most of the males becoming azoospermic by Week 7. The males in the low dose group also exhibited a significant decrease in sperm counts at this week (Fig. 7). Additionally, there was a significant increase in abnormal sperm morphology at Week 7 and some depression in motility. The sperm head distortions observed (Fig. 8) have not been reported in the literature. Partial recovery was apparent in the semen analyses by Week 14, as evidenced by an increase in sperm counts, and further supported by epididymal and testicular histological assessments at Week 16.

Another example of the utility of this model is reflected in an investigation wherein unscheduled DNA synthesis was measured in ejaculates following treatment with methylmethane sulfonate. Rats received an intratesticular injection of ^{3}H-d-thymidine followed immediately by an injection of MMS (50 mg/kg, i.p.). Semen samples were then collected weekly for 10 weeks. The sperm heads were sheared from the tails by sonication; the heads were then separated by centrifugation. Radioactivity in the heads was determined by liquid scintillation counting. In controls, radioactivity was only

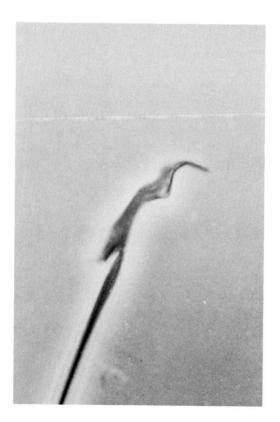

Fig. 8.

detected at 8-10 weeks post-exposure, which corresponds to cells that would have been mitotically-dividing spermatogonia 8-10 weeks earlier (Fig. 9). MMS-treated animals also showed elevated levels of activity at these times. However, MMS also produced an increased incorporation of label between Weeks 4-6 (early spermatid stage). This study serves to illustrate the point that the use of this model allows one to obtain corollary information on the germ cell without compromising the animal. The fact that UDS does not follow identical time courses across animals reinforces the utility of being able to follow an individual animal over time.

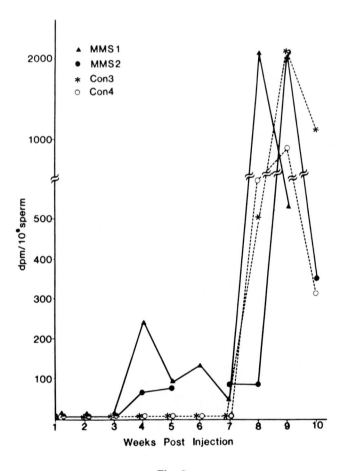

Fig. 9.

References

Ammann, R.P. (1982). Use of animal models for detecting specific alterations in reproduction. Fund. Appl. Tox., 2, 13-26.

Bryan, J.H.D. (1970). An eosin-fast green-naphthol yellow mixture for differential staining of cytologic components in mammalian spermatozoa. Stain. Tech. 45, 231-236.

Chester, R.V. and Zucker, I. (1970). Influence of male copulatory behavior on sperm transport, pregnancy, and pseudopregnancy in female rats. Physiol. Behav. 5, 35-43.

EPA General Toxicology Report (1983). An evaluation of mouse sperm morphology test and other sperm tests in nonhuman mammals. Mutat. Res. 115, 1-72.

Katz, D.F. and Overstreet, J.W. (1981). Sperm motility assessment by video-micrography. Fertil. Steril. 35, 188-191.

Saksena, S.K., Lau, I.F., Chang, M.C. (1979). Age dependent changes in the sperm population and fertility in the male rat. Exptl. Aging Res. 5, 373-381.

Whorton, D., Milby, T.H., Krauss, R.M. and Stubbs, H.A. (1979). Testicular function in DBCP exposed pesticide workers. JOM 21, 161-166.

Footnotes

[1] This research was supported in part by NIOSH Grant R01-OH1271 and EPA Grant CR-808880.

THE MOUSE AS A MODEL SYSTEM FOR MUTATION TESTING AND EVALUATION OF RISK IN MAMMALS

Susan E. Lewis

Research Triangle Institute, P. O. Box 12194, Research Triangle Park, NC 27709

The mouse is a valuable experimental mammal for many kinds of genetic studies, and it provides unique advantages to mutagenesis experiments in particular. Mice have a relatively short generation time, small size for convenient housing, and reproduce readily in the laboratory. Another advantage lies in the wealth of knowledge which has been acquired about the mouse genome. The linkage map of the mouse is fairly well detailed, and a large number of genetically defined inbred strains with a variety of characteristics are available.

The considerable homology of linkage of some mouse loci to those in humans (Lalley et al., 1978; Johnson et al., 1981a) makes extrapolation from mouse results to estimations of human risk seem feasible. Many human genetic disorders have parallels in the mouse. Among these are mouse models of lysosomal storage diseases (Duchan et al., 1980; Miyawaki et al., 1982; Kobayashi et al., 1982; Pentchev et al., 1980). Two X-linked mouse disorders appear to be closely analogous to X-linked diseases in man (Rowe et al., 1977; Prins and Van Hamer, 1980). A mouse model for osteogenesis imperfecta has been recovered following mutagenesis (Guenet et al., 1982). A β-glucuronidase deficiency found in C3H mice results in the accumulation of a metabolite, although no disfunction is detectable (Yatziv et al., 1978). Both α and β-thalassemias have been detected in mice (Russell et al, 1976; Lewis et al, 1983).

© 1984 Alan R. Liss, Inc.

The genetic variation found in mice has led to a number of mutation test systems. The first and most widely used so far is the seven locus visible test system (Russell, 1951) which has given rise to most of what is known about the in vivo induction of transmissable germinal mutations in mammals (Searle, 1975). The system employs seven recessive genetic markers which affect the external appearance of the animal. Six alter the "wild type" coat color and one reduces the size of the ear. An important feature of this system is the use of the tightly linked dilute (d) and shortear (se) genes on chromosome 9 to identify deletions. In order to perform the visible specific locus test, mice with the dominant wild type genotype are treated with a mutagenic agent and then mated with animals homozygous for the seven recessive markers. Mutant progeny are then identified by the external appearance caused by the mutation of one of the treated parents' genes to the recessive mutant type. Most of the knowledge gained so far about the in vivo induction of transmissible mutations in mammals has been gained using the visible specific locus test.

A recently developed mutation test system in the mouse is the electrophoretic system originally described by H. V. Malling (Malling and Valcovic, 1977). Like the visible system, the electrophoretic system is based on genetic differences between the parents of the mice in the test generation. Twenty-one genetic markers are examined, of which ten differ between C57Bl/6J and DBA/2J, the strains currently used in this system (Johnson and Lewis, 1981a; Johnson et al., 1981b). For these ten markers the gene products of both parents can be identified, permitting the study of control and treated loci in the same animal. Furthermore, it is possible, to identify null alleles at these loci unmasked by the normal allele. The electrophoretic system was able to detect both induced and naturally occurring mutations in experiments using ethylnitrosourea (ENU) (Johnson and Lewis, 1981a) and procarbazine (Johnson and Lewis, 1981b; Johnson et al., 1981b).

Many dominant visible mutations have been identified at many different loci in the mouse. However, such mutations have not in general been used for mutation detection because of their generally low spontaneous and induced frequencies (Schlager and Dickie, 1967). Two recently developed test systems search for dominant mutations by using more detailed analysis of the animal than can be afforded by simple visual examination. One involves the preparation and examination of the skeleton (Ehling, 1965; Selby and Selby, 1977; Johnson and Lovel, 1983). Another system involves the examination of the eye with a slit lamp, searching for dominant cataracts (Kratochilova, 1982).

Other test systems are currently being used or under development. The relatively unspecific dominant lethal test can, of course, be performed in mice as well as in other experimental mammals (Generoso et al., 1979). Another method studies the induction of male sterility in the test generation along with cytological analysis of meiotic chromosomes to detect heritable translocations (Generoso et al., 1978). A test system utilizing inversions and homozygotes to recover recessive lethal mutations on specific chromosomes is currently being developed (Roderick, 1979).

In addition, a mutation detection method utilizing the analysis of enzyme activities in the progeny of mutagen-treated parents (Feuers and Bishop, 1979). A feasibility study using the mutagen EMS has been performed and a number of variants recovered. More genetic confirmation of these presumed mutants is needed. However, one mutation identified by its low specific activity, in a slightly different system, has been well confirmed genetically (Johnson et al., 1981c). Thus, it may be possible to identify germinal mutations by screening mice for variations in enzyme specific activities.

It has become apparent from the study of mutation in mice that different parts of the genome may have appreciable differences in mutability and mutation rates. A study of the occurrence of visible mutations in a number of different inbred mouse strains has shown that certain loci appear to be more mutable than others (Schlager and Dickie, 1967). Russell and his co-workers have reported that the $\underline{s}$ locus appears to be particularly susceptable to mutation induction by radiation (Russell, 1965), while this locus is one of the most resistant loci to ENU (Russell, 1983).

Although there is not yet enough data to permit firm conclusions about the differential mutability of the electrophoretically expressed loci, certain tendencies seem apparent. Both the Pep-3 and the Mod-1 locus appear to be more susceptable to mutation-induction by ENU than do the other loci studied in the electrophoretic system (Johnson and Lewis, 1983, Lewis and Johnson, 1983). Recent unpublished data suggests that the Idh-1 locus may also be especially sensitive to the mutagenic action of ENU.

The extensive polymorphism in biochemically-expressed loci in both inbred and wild mice (Selander et al., 1965; Womack, 1979) permits comparisons of naturally-occuring and induced mutations at the same loci. Of the M induced mutations recovered in the electrophoretic system, only one could possibly be the same as the naturally occurring alleles (Lewis and Johnson, 1983). Furthermore, an ENU-induced mutation at the Hba locus has caused a his → leu substitution at position 89 (Popp et al., 1983). This position is conserved in the mammalian hemoglobins studied so far (Bunn et al., 1977). Thus, it appears essential to compare the nature of mutagen-induced with naturally occurring variation.

Another aspect of mutagenesis which can be studied fruitfully in vivo is the effect of genetic background on the sensitivity of the genome to mutation. Certain inbred strains clearly have a higher incidence of spontaneous mutation than others (Schlager and Dickie, 1967). There is less information on possible differences in the sensitivities of different strains to induction mutation. Results in the dominant lethal test, in mice of various strains, show differences in induction of dominant lethals with Ethyl Methanosulfonate (Generoso and Russell, 1969). Preliminary evidence suggests that the induction, by ENU, of electrophoretically expressed mutations in spermatogonia, may be more effective in the C57Bl/6J than in DBA/2J males. Additional data needed to confirm this observation is now being accumulated.

Another important aspect of mutagenesis, which can only be studied at present in vivo, is the effect of sex and germ cell stage on rate of mutation induction. The specificity of the action of mutagens on male germ cells is highly variable. Some, like ENU (Russell et al., 1979; Johnson and Lewis, 1981a), are primarily active in spermatogonia. Others, like EMS and MMS, induce mutations in postspermatogonial stages (Russell et al., 1981) only.

Ethylene oxide appears to have a very narrow stage specificity when administered to male mice by inhalation (Generoso, 1983); dominant lethals are induced in mature stored spermatazoa.

Spermatogonia have been the primary germ cell stage studied, particularly because mutagenic effects on spermatogonia will persist for the rest of reproductive life of the individual. Mutations occurring in other germ cell stages can only affect progeny conceived shortly after exposure and cannot appear in progeny at later times. However, given the conditions of industrial and environmental exposures of humans, it is likely that at least some offspring will be conceived during exposure. Thus, it is most important to explore the effects of mutagens in postspermatogonial stages, in order to determine risk to children born during or shortly after the parents' exposure.

The effect of mutagens on the female germ cell has been subjected to much less study than that of the male, at least partly for tactical reasons. Earlier studies with radiation indicate that the early oocyte is relatively resistant to mutagenesis (Russell, 1951; Russell, 1965).

Finally, the results with short-term tests are not always consistent with those obtained in whole animal systems. As reviewed by L. B. Russell and her co-workers (1981), substances which are highly mutagenic in certain short-term tests may not induce mutations in vivo, and vice versa. Thus, at least some mutagens should be tested in the whole animal. In conclusion, however, the best use of in vivo tests is the exploration of locus, strain, sex and germ-like stage-specific effects of mutagens.

ACKNOWLEDGMENTS

The author's program is funded by NIEHS Contract #N01-ES-0-002. We also thank Teresa Erexson for typing the manuscript and Frank Johnson for discussing the concepts involved.

REFERENCES

Bunn HF, Forget BG, Ranney HM (1977). "Human Hemoglobins." Philadelphia: Saunders.

Duchan LW, Eicher EM, Jacobs JM, Scaravilli F, Teixeira F (1980). Hereditary leucodystrophy in the mouse: The new mutant twitcher. Brain 103:695.

Ehling UH (1965). The frequency of x-ray induced dominant mutations affecting the skeleton of mice. Genetics 51:723.

Feuers RJ, Bishop JB (1979). Definition and assessment of ethyl methanesulfonate induced mutations in C57Bl/6J mice. Environ Mutagenesis 1:125.

Generoso WM, Russell WL (1969). Strain and sex variations in the sensitivity of mice to dominant-lethal induction with ethyl methanesulfonate. Mutat Res 8:589.

Generoso WM, Cain KT, Huff SW, Gosslee DG (1978). Heritable-translocation test in mice. Chemical Mutagens:Principle Methods of Their Detection 5:55.

Generoso WM, Cain KT, Krishna M, Huff SW (1979). Genetic lesions induced by chemicals in spermatozoa and spermatids of mice are repaired in the egg. Proc Natl Acad Sci USA 76:435.

Generoso WM, Cumming RB, Bandy JA, Cain KT (1983). Increased dominant-lethal effects due to prolonged exposure of mice to inhaled ethylene oxide. Mutation Research 119:377.

Guenet JL, Stanescu R, Maroteaux P, Stanescu V (1982). Fragilitas ossium (fro): An autosomal recessive mutation in the mouse. In Animal models of inherited metabolic diseases, New York: Alan R. Liss, p. 265.

Johnson FM, Lewis SE (1981a). Electrophoretically detected germinal mutations induced in the mouse by ethylnitrosourea. Proc Natl Acad Sci USA 78:3138.

Johnson, FM, Lewis SE (1981b). Mouse spermatogonia exposed to a high, multiply fractionated dose of a cancer chemotherapeutic drug: Mutation analysis by electrophoresis.

Johnson FM, Lewis SE (1983). The detection of ENU-Induced mutations in mice by electrophoresis and the problem of evaluating the mutation rate increase, Utilization of Mammalian Specific Locus Studies in Hazard Evaluation and Estimation of Genetic Risks, FJ de Serres and W. Sheridan ed, Plenum NY (in press).

Johnson FM, Lovell D (1983). Dominant skeletal mutations are not induced by ethylnitrosourea in mouse spermatogonia. Abstracts of 14th Annual Mtg Environmental Mutagen Society: 184.

Johnson FM, Chasalow F, Hendren RW, Barnett LB, Lewis SE (1981a). A null mutation at the mouse Phosphoglucomutase-1 locus and a new locus, Pgm-3, Biochem Genet 19:599.

Johnson FM, Roberts GT, Sharma RK, Chasalow F, Zweidinger R, Morgen A, Hendren RW, Lewis SE (1981b). The detection of mutants in mice by electrophoresis: Results of a model induction experiment with procarbazine. Genetics 97:113.

Johnson, FM, Chasalow F, Anderson G, MacDougal P, Hendren RW, Lewis SE (1981c). A variation in mouse kidney pyruvate kinase activity determined by a mutant gene on chromosome 9. Genet Res 37:123.

Kobayashi T, Nagara H, Suzuki K, Suzuki K (1982). The twitcher mouse: Determination of genetic states by galactosylceramidase assays on clipped tail. Biochemical Medicine 27:8.

Kratochvilova J (1981). Dominant cataract mutations detected in offspring of gamma-irradiated male mice. J Hered 72:302.

Lalley PA, Minna JD, Francke V (1978). Conservation of autosomal gene synteny groups in mouse and man. Nature 274:160.

Lewis SE, Johnson FM (1983). Dominant and recessive effects of electrophoretically detected specific locus mutations, Utilization of Mammalian Specific Locus Studies in Hazard Evaluation and Estimation of Genetic Risks, FJ de Serres and W. Sheridan ed, Plenum NY (in press).

Lewis SE, Popp RA, Johnson FM and Skow LC (1983). A spontaneously arisen β-thalassemia in mice. Abstracts of the 14th Annual Mtg, Environmental Mutagen Society, p. 181.

Malling HV and Valcovic LR (1977). A biochemical specific locus mutation system in mice. Arch Toxicol 38:45.

Miyawaki S, Mitsuoka S, Sakiyama T, Kitagawa T (1982). Sphingomyelinosis a new mutation in the mouse, A model of Niemann Pick disease in humans. Jour Hered 73:257.

Popp RA, Bailfiff EG, Skow LC, Johnson FM and Lewis SE (1983). Analysis of a mouse α-globin gene mutation induced by ethylnitrosourea, Genetics (in press).

Prins HW, Van den Hamer CJA (1980). Abnormal copper-thionein synthesis and impaired copper utilization in mutated Brindled mice: Model for Menkes' disease. J Nutr 110:151.

Roderick TH (1979). Chromosomal inversions in studies of mammalian mutagenesis. Genetics 92:s121.

Rowe DW, McGoodwin EB, Martin GR, Grahn D (1977). Decreased lysyl oxidase activity in the aneurism-prone, mottled mouse, J Biol Chem 252:939.

Russell LB, Russell WL, Popp RA, Vaughan C, Jacobian KB (1976). Radiation-induced mutations at mouse hemoglobin loci. Proc

Natl Acad Sci USA 73:2843.

Russell LB, Selby PB, vonHalle E, Sheridan W and Valcovic L (1981). The mouse specific-locus test with agents other than radiations. Mutation Research 86:329.

Russell WL (1951). X-ray induced mutations in mice, Cold Spring Harbor Symp. Quant Biol 16:327.

Russell WL (1965). Evidence from mice concerning the nature of the mutation process. In Geerts SJ (ed): "Proc XI Int Congr Genet" (The Hague, 1963) Vol 2, New York: Pergamon Press, p. 257.

Russell WL (1983). Relation of mouse specific locus test to other mutagenicity tests and to risk estimation in Utilization of mammalian specific locus studies in hazard evaluation and estimates of genetic risks. F.J. de Serres and W.S. Sheridan, ed., in press.

Russell WL, Kelly EM, Hunsicker PR, Bangham JW, Maddux SC, Phipps EL (1979). Specific-locus test shows ethylnitrosourea to be the most potent mutagen in the mouse. Proc Natl Acad Sci USA 76:5818.

Schlager G, Dickie MM (1967). Natural mutation rates in the house mouse. Estimates for five specific loci and dominant mutations. Mutation Res 11:89.

Searle AG (1975). The specific locus test in the mouse. Mutation Res 31:277.

Selander RK, Yang SY, Hunt WC (1965). Polymorphism in esterase and hemoglobin in wild populations of the house mouse. Studies in Genetics V:271.

Selby PB, Selby PR (1977). Gamma-ray induced dominant mutations that cause skeletal abnormalities in mice: I. Plan, summary of results, and discussion. Mutation Res 43:357.

Womack J (1979). Single gene difference controlling enzyme properties in the mouse, Genetics 92 Suppl, Proceedings of the Workshop: Methods in Mammalian Mutagenesis:85.

Yatziv S, Erickson RP, Sandma R, Robertson WVB (1978). Glycosaminoglycan accumulation with partial deficiency of β-glucuronidase in the C3H strain of mice. Biochem Genet 16:1079.

CHILDHOOD TUMORS AND PARENTAL OCCUPATIONAL EXPOSURES

John M. Peters, M.D.
Susan Preston-Martin, Ph.D.
University of Southern California
School of Medicine
Department of Family and Preventive Medicine
Los Angeles, California   90033

What we plan to present is a study that looks at cancer in children related to parents' occupations. We would like to describe the results of a preliminary study (1) and then describe studies that would follow-up on this preliminary study. In Los Angeles County we have a population-based tumor registry covering about eight million people which generates over 25,000 new cases of cancer each year. Information is abstracted on nearly all of the new tumors. From this collection of cases which began in 1972, it is possible to do all kinds of studies related to cancer. We collected information on 92 brain tumors in children under age 10. The cases were matched to a friend or neighborhood control. The mothers of the cases and controls were interviewed. Along with other information, an occupational history was taken from the mothers on both parents. This information was computerized. We then looked at the specific jobs, the specific exposures and the specific industries. The questionnaire that was used asked mothers about whether they got chemicals on their skin or inhaled them. The mother told us about the father's occupational exposures. We also went through all of the father's jobs and determined whether there was solvent exposure and paint exposure. Table 1 summarizes the results. Three times as many mothers of cases inhaled or got chemicals on their skin. For fathers, jobs involving exposure to solvents carried a risk of about three and for a subdivision of that (paint) the risk was seven even though the numbers are small. Probably the most interesting finding is the fact that fathers of cases were much more likely to work in the aircraft industry than the fathers of controls. In this

© 1984 Alan R. Liss, Inc.

case the risk was 10-0. We further looked at those 10
individuals (all of the people that worked in the aircraft
industry) to see if there were any possible hints on specific
exposure. This is presented in Table 2. This sort of
analysis did not reveal anything definitive, although
trichloroethylene was mentioned twice. The other jobs,
even though not mentioning solvents per se, may have involved
solvent exposure. We would summarize this particular
study by saying that chemical exposure to mothers and
fathers increases the risk for childhood brain tumor. We
do not really know what the specific agent(s) might be.
In an attempt to see when exposure might be critical, that
is exposure prior to pregnancy, exposure during pregnancy
or exposure after pregnancy; we were not able to say because,
in general, the men had the same job at all three points.

The follow-up effort which we will describe has the
goal of identifying specific exposures that might be responsible for the increased risk.

Ways in which maternal exposure could be transmitted
are summarized in Table 3. Table 4 contains ways in which
paternal exposure could result in risk to child. All of
these possibilities must be considered in trying to pursue
possible specific agents.

Table 5 presents the sequence of events necessary to
conduct the follow-up study. With our Cancer Surveillance
Program (CSP) for Los Angeles County we get the reported
tumors coming in to the registry. From that we identify
the case. We then obtain permission from the case's
physician to contact the parents. That results in about a
95% permission rate. We then ask the parents if they
would allow us to interview them. They permit interviews
in about 95% of the cases. In the study that we are now
doing, we are interviewing both the mother and father.

The questionnaire contents are presented in Table 6.
We are interested in birthdates, ethnicity, marital status,
religion and education, and other demographic variables.
Radiation history is important for both the child and both
parents. Pregnancy history including durations, use of
medicines, complications, etc., will be collected. Family
history of malignancy or congenital malformations will be
elicited as well as personal habits of both the parents
and the child; alcohol, tobacco and drug use, plus a complete

occupational history on both parents. Food consumption patterns for both parents and child will also be attained.

Table 7 outlines the elements of the occupational history. We want to know where the people worked, what their job title was, and what materials they handled. We also want some type of description of the job tasks and exposures. In addition to that (see Table 8), we ask specific questions about the skin and clothes contamination, the use of protective clothing, some qualitative estimate on the part of the individual about their degree of inhalation; that is, how bad are the working conditions for gas and vapors, fumes or dust. We also want to know whether they use respiratory protection or other forms of protection on the job. In addition, we are determining whether radiation exposure occurred on the job.

Table 9 contains a summary of some of the substances that will be asked about specifically. We will specifically ask them about the four most commonly used plastics. We will give them two or three examples from each of the major solvent categories. We will continue through the categories in this fashion.

In summary, we have tried to present the findings that suggest a possible relationship between parents' exposure and childhood tumors. It obviously needs to be followed up and we have outlined how we expect to do that. We hope two or three years from now at a meeting similar to this we will have some results to talk about that can lead directly to some preventive action that will reduce the risk of childhood tumors.

TABLE 1

Table 1. Matched-pair comparison of parental occupational exposure of cases and controls. The mother was considered to have been exposed if exposed at any time from 1 year before conception through lactation. The father was considered to have been exposed if exposed during that period or at the time of diagnosis of a case.

Factor	Concordant pairs: both exposed	Discordant pairs		Relative risk	One sided test (P)
		Cases exposed	Controls exposed		
Mother					
Got chemicals on skin	0	10	3	3.3	.05
Inhaled chemicals or fumes	1	12	4	3.0	.04
One or both of the above	1	14	5	2.8	.03
Father					
Exposed to chemical solvents	3	17	6	2.8	.02
Exposed to paints	0	7	1	7.0	.04
Worked in aircraft industry	2	10	0	∞	.001

Table 2. Occupational information on parents of cases and controls who worked in the aircraft industry.

Diagnosis	Child's age	Mother's occupation	Father's occupation	Exposure
Astrocytoma	9	Wire soldering*†	Electronics assembler*†‡	Solder
Oligodendroglioma	8		Machinist*†‡	Trichloroethylene
Astrocytoma	6		Machinist*†‡	Dust, oils
Astrocytoma	7		Production scheduling and parts inspection*†	Trichloroethylene, methyl ethyl ketone
Medulloblastoma	4		Electrical engineer*†=	
Glioma	4	Secretary*		
Astrocytoma	4	Keypunch operator*†	Computer operator*†‡	
Medulloblastoma	<1		Scientist-physicist*†‡	Ionizing radiation (wore film badge)
Astrocytoma	<1		Plane painter*†‡	Spray paint
Astrocytoma	9		Engineer‡	
Astrocytoma	8		Wing parts inspector*†	
Medulloblastoma§	4	Stockroom clerk*†		Dust, exhaust
Control		Secretary*†		
Medulloblastoma§	3		Flight line mechanic*†‡	Exhaust
Control			Electrical engineer*†=	
Medulloblastoma§	<1		Student-aircraft mechanic school*†‡	
Control		Secretary*	Technician-student*†‡	

*Occupation at any time during the year before pregnancy. †Occupation at any time during the pregnancy. ‡Occupation at time of diagnosis. §A parent of the case and matched control both had a history of employment in the aircraft industry.

TABLE 3

SEQUENCE OF STUDY

- Identify Case From CSP.

- Obtain Permission From Case's M.D. To Contact Parents To Explain Study.

- Obtain Parents' Consent For Interview.

- Interview Case Mother.

  • Get Occupational History On Father and Mother From Mother.

- Interview Case Father If Available (Expect 80%).

- Select Friend or Neighborhood Control(s) (Walk Algorithm).

- Identify Control Mother.

- Interview Control Mother.

- Interview Control Father.

TABLE 4

MOTHER'S OCCUPATIONAL EXPOSURE

Exposure To The Mother Can Lead To Effects In The Child In Several Potential Ways:

- Maternal Exposures Prior To Pregnancy Could Result In Transmissible Genetic Effects That Could Cause Childhood Neoplasia.

- Maternal Exposures During Pregnancy Could Result In *In Utero* Exposure To The Developing Infant.

- Maternal Exposures Prenatally or During The Neonatal Period Could Result In Transmission Of Exposures In Breast Milk.

- Maternal Exposures Following Birth Can Be Brought Home In The Form Of Soiled Clothing To Which The Child Can Be Directly Exposed.

TABLE 5

FATHER'S OCCUPATIONAL EXPOSURE

Exposure To The Father Could Affect The Child In Several Potential Ways:

- Paternal Exposures Prior To Pregnancy Could Result In Transmissible Genetic Effects That Could Result In Childhood Neoplasia.

- Paternal Exposures Brought Home On Clothing, For Example, Could:

    (1) Expose The Mother And Produce Genetic Effects.

    (2) Expose The Pregnant Mother And Affect The Child In Utero.

    (3) Expose The Pregnant Or Postnatal Mother And Affect The Child Through Breast Milk, Or

    (4) Expose The Newborn Directly.

TABLE 6

INFORMATION TO BE COLLECTED FROM PARENTS

- Demographic And Identifying Information For Child And Both Parents (Address, Birthdates, Ethnicity, Marital Status, Religion, And Education).

- Radiation History Of Child And Both Parents (Diagnostic X-Rays, Radiation Treatment, Etc.).

- Pregnancy History (Duration, Use Of Medicines, Complications, Etc.).

- Family History Of Malignancy Or Congenital Malformations.

- Personal Habits Of Both Parents And Child (Alcohol, Tobacco, Drug Use).

- Occupational History (Complete For Both Parents).

- Food Consumption Patterns For Both Parents And Child.

TABLE 7

The Questionnaires Will Contain A Full Occupational History Consisting Of The Following Elements For Each Job:

(1) Employer

(2) Job Title

(3) Materials Handled

(4) Job Tasks And Exposures

TABLE 8

INFORMATION TO BE COLLECTED FOR EACH JOB

(1) Skin And Clothes Contamination

(2) Inhalation Of Chemical Gases, Vapors Or Fumes

(3) Inhalation Of Dust

(4) Use of Respirators

(5) Radiation Exposure

TABLE 9

EXAMPLES OF CATEGORIES AND SPECIFIC SUBSTANCES TO WHICH PARENTS MIGHT BE EXPOSED

- Solvents
- Plastics
- Dyes
- Inks And Pigments
- Paints
- Fuels
- Formaldehyde
- Oils
- Inorganic Dusts
- Polycyclicaromatic Hydrocarbons
- Acids
- Alkalis
- Pesticides
- Synthetic Textiles
- Metals
- Organic Dusts
- Rubber Chemicals
- Welding Fumes

REFERENCES

1. Peters JM, Preston-Martin S, Yu MC (1981). Brain tumors in children and occupational exposure of parents. Science 213:235.

MALE-TRANSMITTED DEVELOPMENTAL AND NEUROBEHAVIORAL DEFICITS

P.M. Adams,[1,2] O. Shabrawy,[3] M.S. Legator[3]

[1] Dept. Psych. & Behav. Sci., [2] Dept. Pharm. & Tox., [3] Dept. Preven. Med. & Comm. Health. The Univ. Texas Med. Branch, Galv. TX.

Birth defects are a result of reproductive dysfunction which can arise from either genetic alterations of the gametes or from insult to the fetus during gestation. While considerable attention has been directed toward the study of morphological evidence of reproductive dysfunction relatively little research has been done on the identification of deficits in behavioral development. In humans this evidence is difficult to identify early because of the relatively slow development of many motor and cognitive behaviors.

In contrast, animal studies of the effect of drug exposure during gestation on the behavioral development of the offspring have increased dramatically in recent years. The identification of behavioral anomalies or developmental differences in the absence of morphological signs of teratogenicity has led to the use of the term behavioral teratogens to describe these in utero drug effects. While higher dosages of these drugs given in utero frequently produce morphological signs of teratogenicity, it is the greater sensitivity of the behavioral assessment for detecting teratogenic effects that is important in making decisions concerning reproductive risk.

While the number of studies on the effects of maternal exposure to drugs or chemicals on the behavioral development of the offspring are

© 1984 Alan R. Liss, Inc.

substantial, the amount of research on the effects of paternal exposure to a drug or chemical on the behavioral development of the subsequent offspring is sparse.

Perhaps the most thoroughly studied behavioral teratogen is alcohol and while the role of maternal alcohol consumption in producing fetal alcohol syndrome is of major importance the role of paternal alcohol use has also been acknowledged (Anderson, 1982). The extent of reproductive toxicity in the male from alcohol includes: loss of libido, reduced sperm count, infertility problems and subtle genetic effects. Evaluations of the offspring of alcohol-consuming male rodents indicated abnormally high mortality rates, retarded growth rates, increased physical deformities, reduced litter size (specifically, fewer females) and behavioral impairments (Klassan and Persaud, 1976; Nice, 1917; Pfeifer et al., 1977; Stockard and Papanicolaou, 1916). Studies on the offspring of men who reported heavy consumption of alcohol at the time of conception suggested that morphological anomolies were present despite a reported absence of drinking by the mother (Bartoshesky et al., 1979; Scheiner et al., 1979).

The effects of paternal lead exposure on the behavior of the offspring was studied in rats by Brady et al. (1975). They found the offspring of lead-exposed male rats were significantly different on a T-maze water discrimination task when tested at thirty days of age. The offspring of the exposed-male rats made more errors than the controls and as many as the females exposed preconception and during pregnancy. These findings strongly suggested a genotoxic effect of lead which was detectable by behavioral assessment of the offspring.

A number of drugs have been reported to have adverse effects on the offspring of exposed male parents including: caffeine, morphine, methadone, and marijuana (Joffe, 1979). The present work is part of a project designed to establish the use of behavioral assessments on the developing offspring of exposed male parents as a sensitive measure of genotoxicity. Our initial experiments utilized the well-studied drug cyclophosphamide (cytoxan) because so much is known about its mutagenic properties and dose-response characteristics in other assays of genotoxicity.

We have chosen the rat for this research and specifically the F344 inbred strain. An inbred strain was chosen in order to reduce genetic heterogeneity. Since we anticipated subtle effects on the behavioral development of the offspring of exposed male rats it was important to reduce the variability of the animals both within and across litters. Therefore, we used brother-sister matings of animals reared from our own colony which was in the ninth generation at the start of the study.

In order to maximize the liklihood of detecting an effect on behavioral development we used an extensive testing battery as shown in Table 1. The battery included

TEST	AGE AT TESTING (DAYS)
Surface Righting	3 - 6
Cliff Avoidance	4 - 10
Swimming	6, 8, 10, 12, 14
Open Field Activity	14, 21
One Way Active Avoidance	30

Table 1. Testing Battery for Assessing Behavioral Development in the Rat.

assessments of simple reflexes (e.g., surface right), motor coordination (e.g., swimming), locomotor activity, and a relatively easy learning task (e.g. active avoidance). The age at which each of these behaviors was assessed is shown in Table 1. These behaviors develop quite rapidly and reach essentially an asymptotic level at an early age. This allows for possible early detection of a genotoxic effect resulting from paternal exposure. A brief description of each of the behaviors assessed and some normative data are given below.

<u>Surface Righting</u>. Is measured by placing the animal on its back on a flat surface and recording the amount of time the animal requires to roll over so that it rests on its stomach and four feet. This behavior develops rapidly and very few animals fail to perform this behavior by 6 days of age. To date we have found this behavior to be insensitive to the detection of genotoxic effects from paternal exposure.

<u>Cliff Avoidance</u>. Is measured by placing the neonate on an elevated platform with the forepaws and fore portion of the head extended beyond the edge. Upon release of the animal, the time required to completely retract the head from the edge of the platform is recorded. A maximum of 60 seconds is allowed. A criterion of twenty seconds was used for successful completion of the head retraction. The development of this behavior is shown in Figure 1.

<u>Swimming Behavior</u>. Develops rapidly in the rat as illustrated in Figure 2. Ratings of the direction of swimming (2=circular movement, 3=straight line movement) are made on each day of measurement. In addition, head angle in the water is measured as: 1=top of head only above the surface, 2=nose and top of head above, 3=ears ½ out of water, 4=ears completely out of the water. Limb movement is rated as either 1=all 4 limbs used, or 2=back limbs only used. Latency to begin limb movement after being placed in the water is also recorded.

<u>Activity</u>. Is measured on an open-field consisting of an elevated platform 45cm in diameter. It was marked off in quadrants with an equal number of areas in each quadrant. The number of entries the animal made during a one minute trial was recorded 3 times with 30 seconds between trials. Activity level on day 14 and 21 is shown for each of the trials in Figure 3.

<u>Active Avoidance</u>. Is measured in a standard

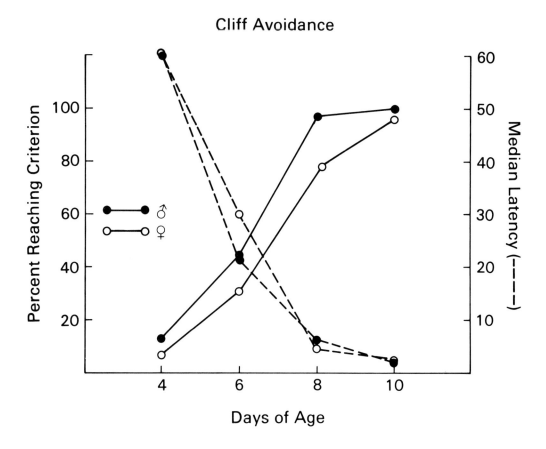

Figure 1. Development of cliff avoidance behavior in the F344 rat.

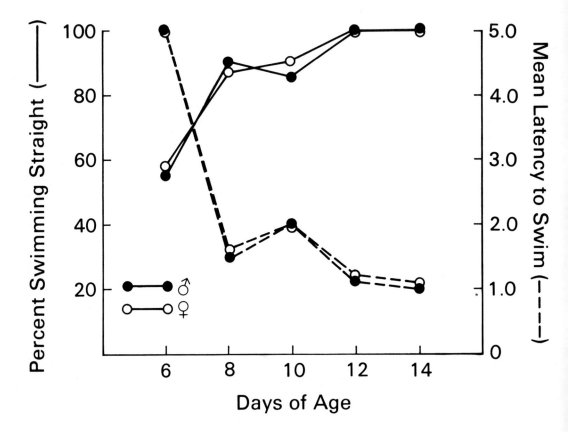

Figure 2. Swimming direction development and latency to begin limb movements as a function of age.

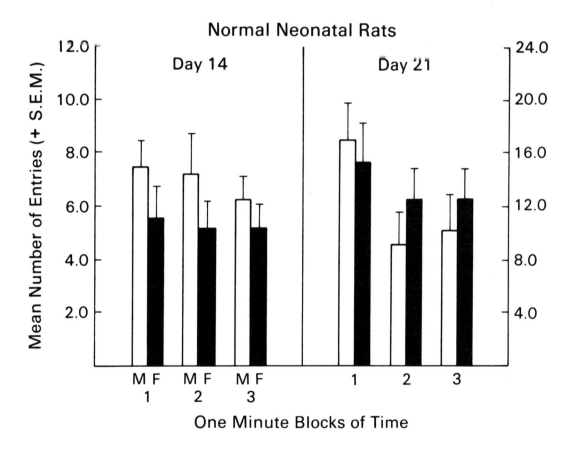

Figure 3. Open-field activity in the rat before (14) and after (21) eye-opening.

operant conditioning chamber equipped with a grid floor. There is a Plexiglas platform against the back wall of the chamber onto which the animal learns to climb in order to escape or avoid electric shock (.5MA). The rat is habituated to the chamber for one minute, removed and placed in a holding cage for 20 seconds. The rat is then placed into the chamber facing away from the platform. Seven seconds after placement, the grid is electrified for a maximum of 60 seconds or until the rat has escaped by climbing onto the platform. The animal is removed from the chamber after each trial and given a total of 15 acquisition trials. Latency to escape or avoid are recorded and a criterion for acquisition is 4 avoidance responses out of 5 successive trials. In Figure 4 is shown typical acquisition data for male and female 30 day old F344 rats. Acquisition criterion is usually met between 7-9 trials and there are no significant differences between males and females. Twenty-four hours later performance was assessed with no shock presented regardless of whether the animals avoided. A criterion for extinction consisted of 12 non-avoidance responses out of 15 successive trials. Avoidance performance is shown in Figure 5 and a gradual increase in response latency is observed for both males and females.

<u>Paternal Cyclophosphamide Exposure</u>. Our first series of experiments involved chronic treatment of males with cyclophosphamide (10mg/kg x 5 days/5 weeks) followed by mating with either CP-treated or saline-treated females (Adams et al., 1981). The dosage selected was sufficiently low to allow the males to remain fertile and to produce minimal dominant lethality. However, females exposed to this dosage regime were found to have reduced fertility and reduced litter size. If we consider those offspring from the breeding of CP-exposed males with saline-exposed females compared to the offspring of a saline-male/saline-female cross the following results were observed. There was a slower rate of development of both cliff avoidance reflexive behavior and swimming ability as reflected in fewer animals swimming straight through postnatal day 10. Open field locomotor activity

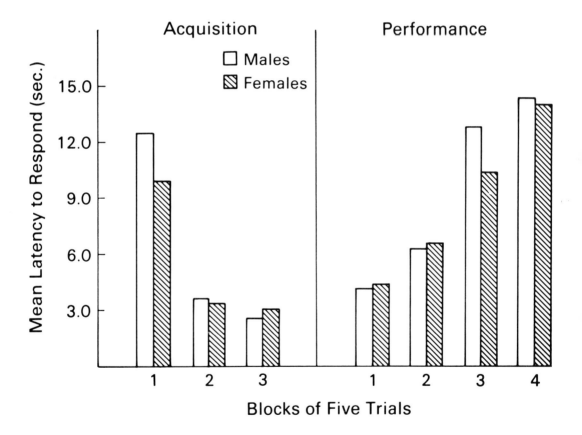

Figure 4. Acquisition and extinction of one-way active avoidance behavior in the 30 day old rats.

was significantly greater in the CP progeny on both days 14 and 21.

Active avoidance behavior was found to show no significant differences in acquisition for either males or females (Adams et al., 1982). Performance of avoidance behavior 24 hours after acquisition was significantly different (Figure 5). The CP-offspring made significantly more avoidance responses regardless of gender than the offspring of the saline-treated males. This perseveration of avoidance responding has been previously shown to be associated with lesions to the frontal cortex and a general state of behavior disinhibition.

The difficulty with a chronic regimen as used in the above experiment is the inability to determine whether one stage of spermatogenesis is more sensitive to the chemical and thus, perhaps more involved in the transmission of the genotoxic effects to the F_1 progeny. In order to determine whether the behavioral anomalies would allow the identification of a genotoxic insult to specific stages of spermatogenesis the following study was conducted (Fabricant et al., 1981). Exposed male rats were bred with untreated females at selected intervals following CP or saline treatment (a single injection). Breeding times were selected to detect genotoxicity to mature sperm (7-9 day breeding), spermatids (14-16 day breeding) or spermatocytes (28-30 day breeding).

The F_1 progeny derived from these breeding times were assessed for evidence of behavioral anomalies. Cliff avoidance development was only slowed in the progeny of the 14-16 day breeding. These neonates were significantly slower on days 6 and 8 but reached a comparable level by day 10 (see Figure 6). A similar developmental lag was observed on swimming behavior (see Table 2). The postmeiotic breeding groups (7-9 and 14-16) have been combined as there were no differences between them. The progeny of the postmeiotic breedings were slower to develop swimming ability but were eventually able to attain a comparable level of performance by 14 days of age.

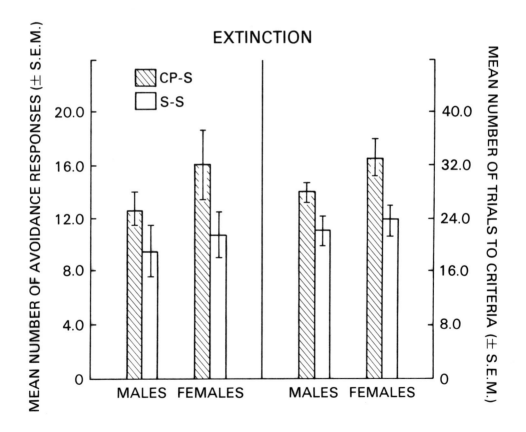

Figure 5. The effect of chronic paternal CP exposure on the performance of active avoidance by the F_1 progeny.

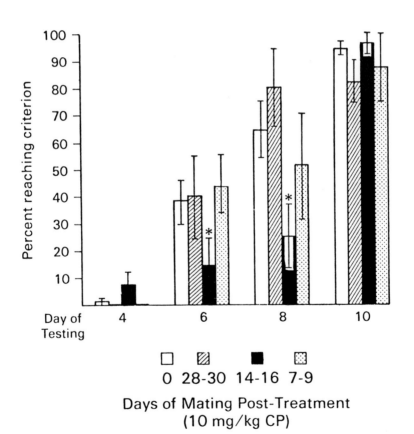

Figure 6. The effect of acute paternal exposure to CP on cliff avoidance development in the F_1 progeny.

	TREATMENT GROUPS		
	Saline	Pre-meiotic	Post-meiotic
N*	6 (59)	4 (32)	6 (51)
Age of Progeny (days)			
6	40.9±5.3	53.0±3.0	25.2±7.3
8	70.8±8.5	68.5±4.9	18.7±5.2**
10	85.4±4.7	79.2±7.2	47.3±11.8**
12	100.0±0.0	100.0±0.0	78.8±12.0
14	100.0±0.0	100.0±0.0	91.7±8.3

N* Indicates number of litters analyzed and the total number of pups tested.
** Statistically significant $p < 0.05$.

Table 2. Percentage of F_1 Progeny from CP-Treated Males and Controls Capable of Swimming in Straight Lines.

Open field activity was found to be significantly different on postnatal day 14 for the 14-16 day breeding interval. In contrast to the progeny of chronic CP-exposed males, these animals were less active than the offspring of the other treatment groups. While it conforms to the findings of the other behavioral endpoints as reflecting an effect on the postmeiotic stage of spermatogenesis the direction of the change is the opposite to that predicted from the chronic CP study. These directional differences could reflect the differences in total dosage and/or in the extent of genetic damage responsible for this particular behavior change.

Active avoidance acquisition was not significantly altered by CP exposure at any of the post-treatment breeding times. Performance, however, was significantly different in the 14-16 day breeding group as reflected by trials to reach the extension criterion (Figure 7).

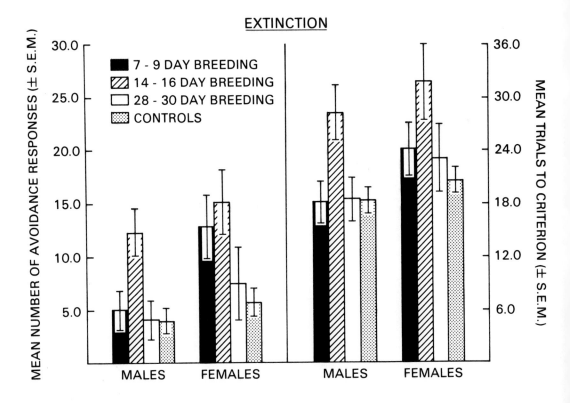

Figure 7. Active avoidance performance of the F_1 progeny of male rats given acute exposure to CP prior to breeding.

In summary, these initial experiments clearly
demonstrated the usefulness of behavioral
assessments of the F_1 offspring to detect effects
of paternal exposure to a genotoxic agent.
Further, these studies demonstrated the ability of
the behavioral endpoints to differentiate
postmeiotic genetic toxicity from premeiotic effects
of cyclophosphamide. This does not mean that
premeiotic genetic damage does not occur but only
that for cyclophosphamide only postmeiotic effects
were detected with the behavioral assessments we
have used.

In order to determine if behavioral assessments
would be useful in detecting the transmittance of
genotoxic effects to subsequent generations the F_1
offspring were bred at 90-100 days of age in
brother x sister crosses. The F_2 offspring of the
chronic and acute F_1 animals were then evaluated
for behavioral anomolies.

The development of swimming behavior in the
F_2 progeny of chronic CP-treated grandfathers was
slower to progress than the F_2 progeny of
saline-treated grandfathers (Figure 8). The impaired
development was most apparent on days 6-10 as it
was in their parents (F_1). The F_2 offspring of
the premeiotic and postmeiotic F_1 offspring
demonstrated an effect comparable to that of the
F_1 animals (Figure 9). The postmeiotic F_2 animals
were slower to develop though not to the same
extent as the chronic F_2 or their parents (F_1).

Open field activity in the F_2 offspring of the
chronic CP treated males was not significantly
different from the control F_2 offspring on day 14.
However, on day 21 these same F_2 animals were
hypoactive relative to the control F_2 group. This
hypoactivity in the F_2 animals is the opposite of
the hyperactivity displayed by their F_1 parents.
The F_2 offspring of the premeiotic and postmeiotic
breeding time were not significantly different from
control F_2 animals on day 14. On day 21, the 14-16
day F_2 offspring were hypoactive relative to the
7-9 and saline F_2 offspring.

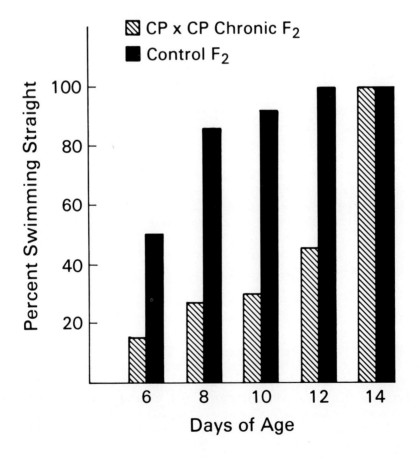

Figure 8. Swimming development of the F_2 progeny derived from chronic CP exposed F_0 male rats.

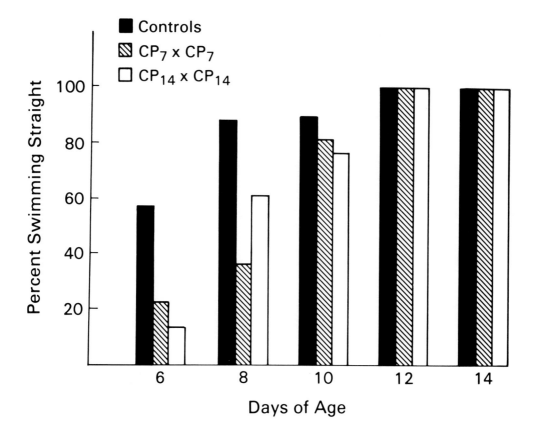

Figure 9. Swimming development of the F_2 progeny derived from acute CP exposed F_0 male rats.

Active avoidance acquisition was not significantly different for the F_2 offspring of either the chronic or acute CP treatment groups compared to the controls. Performance of the avoidance behavior was found to be significantly altered in the F_2 of the chronic CP group with the CP F_2 animals persisting in avoidance behavior as shown by their F_1 parents as shown in Figure 9. A similar finding was observed in the CP 14-16 F_2 offspring comparable to the performance difference observed in the F_1 animals of this breeding interval.

In summary we feel we have demonstrated quite clearly that behavioral assessment of the developing offspring from an exposed male are capable of detecting genotoxicity. Furthermore, that through selective breeding to isolate specific stages of spermatogenesis, the behavioral assessments of these offspring will be sensitive to a genotoxic effect during that particular stage unless it undergoes repair prior to breeding with untreated females. Finally, from our F_2 data it appears these behavioral assessments are sufficiently sensitive to detect genotoxic effects that are transmitted to the second generation. While much more needs to be done with this assessment approach, particularly with regard to the evaluation of other drugs and chemicals, the use of behavioral endpoints to detect genotoxicity appears to be a viable tool in the arsenal of the toxicologist.

REFERENCES

Adams, PM, Fabricant, JD, Legator, MS (1981). Cyclophosphamide induced spermatogenic effects detected in the F_1 generation by behavioral testing. Science 211:80-83.

Adams, PM, Fabricant, JD, Legator, MS (1982). Active avoidance behavior in the F_1 progeny of male rats exposed to cyclophosphamide prior to fertilization. Neurobehav Toxicol Terat 4:531-534.

Anderton, RA (1982). The possible role of paternal alcohol consumption in the etiology of the fetal alcohol syndrome. In Abel, EL (ed): "Fetal

Alcohol Syndrome Vol III Animal Studies,"
Florida: CRC Press
Bartoshesky, LE, Feingold, M, Scheiner, AP, Donovan, CM (1979). A paternal fetal alcohol syndrome and fetal alcohol syndrome in fetal alcohol syndrome in a child whose alcoholic parents had stopped drinking. Birth Defects Conf.
Brady, K, Herrera, Y, Zenick, H (1975). Influence of paternal lead exposure on subsequent learning ability of offspring. Pharm Biochem Behav 3:561-565.
Fabricant, JD, Legator, MS, Adams, PM (1981). Post-meiotic germ cell mediation of behavioral anomalies in progeny of males treated with cyclophosphamide. Environmental Mutagenesis Society Abstracts.
Joffe, J (1979). Influence of drug exposure of the father on perinatal outcome. Clin Perinatal 6:21.
Klassen, RW, Persaud, T (1976). Experimental studies on the influence of male alcoholism on pregnancy and progeny. Exp Path 12:38.
Malakhouskii, VG and Prozorovskii, VB (1975). Behavioral disorders in the progeny of rats subjected to chronic action of alcohol. Farm Toksik 38:88.
Nice, LB (1917). Further observations on the effects of alcohol on white mice. Am Nat 51:596.
Pfeifer, WD, Mackinnon, JR, Seiser, RL (1977). Adverse effects of paternal alcohol consumption on offspring in the rat. Bull Psychonom Soc 10:246.
Scheiner, Ap, Donovan, CM, Bartoshesky, LE (1979). Fetal alcohol syndrome in child whose parents had stopped drinking. Lancet 1:1077.
Stockard, CR, Papanicolaou, G (1916). A further analysis of the heredity transmission of degeneracy and deformities by the decendents of alcoholized mammals I. Am Nat 50:65.

QUESTIONS AND ANSWERS

Question. Is there any possibility that you have some selective phenomena here? Your litter size is reduced to one-half. What kind of selection might go on?

Answer. The litter size was only reduced to ½ in the offspring of the chronic cyclophosphamide exposed males. In the acute study the litter size was normal. It is quite possible that the viable embryos making it through to term and then evaluated in the behavioral studies are the offspring least affected by the genotoxic insult. This is a problem frequently sited in studies of in utero drug exposure protocols. However, it is more of a problem when one tries to explain results on the viable offspring in the presence of evidence of fetal lethality than in the present case.

Question. I have a couple of questions about the F_1 offspring. First of all it was not clear to me whether the time of eye opening and cliff avoidance were correlated.

Answer. The F344 strain rat typically does not open its eyes until approximately 16 days of age. We assess cliff avoidance between day 4 and day 10 so none of the animals have their eyes open at the time of testing.

Question. If an animal was defective by some criteria on one of these behaviors was it also defective on one of the other behaviors?

Answer. Not necessarily all the other behaviors but usually more than one.

Question. If you take one of these animals and mate them to an untreated animal in a backcross fashion do you see evidence to suggest that there is single dominance?

Answer. We have done some backcross breeding but we don't have enough litters at the present time to answer the question.

Question. Are you surprised that there were not gross differences between the groups since body weight, for example, is often found to be sensitive?

Answer. I was fully prepared to be doing analysis of covariance statistics on much of these data but so far that has not been necessary. Certainly we find in utero exposure to a drug frequently produces birth weight and growth rate differences. Perhaps such differences are less of a problem in preconception male exposure protocols.

Question. In your genetic transmission experiment how did you do the breeding of the F_2 progeny?

Answer. These were brother x sister matings. So with each treatment the F_1 progeny of a given litter were inbred. All of the F_2 progeny were then tested in the same manner as the F_1 animals.

Question. Is there any difference between the F_1 and F_2 animals with respect to the observed difference between control and experimental treatment?

Answer. I cannot answer that question completely because there are additional statistical analyses that need to be done. However, the F_2 progeny of an experimental treatment were more like F_2 controls for some behaviors (e.g. cliff avoidance). I think it will also be borne out by further analysis that the acute F_2 progeny are more like the controls than are the chronic F_2 progeny. Some behaviors may prove to be more likely to show this regression in the F_2 offspring than others. This requires consideravle analysis at both the across litter and within litter level before a conclusion is reached.

Index

Abortion, spontaneous
   and DBCP, 25–26, 38–39, 41–43
   and induced sperm changes, 99
   monitoring early fetal loss, 20
Acetaldehyde, nicotine, and caffeine in rats, 52
Age
   confounding variable, sperm tests, 96
   and maternal risk, 48
Aircraft industry and childhood tumors, 137–138, 140–141; *see also* Childhood tumors and parental occupational exposure
Alcohol
   male-transmitted effects, 150
   and maternal risk, 46
   sperm tests as indicators of germ cell damage, 92, 97
   and testicular function, 46
   see also Fetal alcohol syndrome
Animal models/studies
   acetaldehyde, nicotine, and caffeine in rats, 52
   fertility as measure of reproductive toxicology, 61, 64
   and risk assessment, 4, 6
   sperm changes, rule, 100
   in surveillance, toxicology, 16–18
   zona-free hamster eggs, 77–78
   see also Cyclophosphamide, male-transmitted effects in rats; Male reproductive function evaluation/tests; Mutation testing and evaluation, mouse model
Appalachia, neural tube defects, 13, 18
Asbestos, 19
Astrocytoma, 141
Atlanta, Congenital Defects Program, 12, 18
Automated image analysis, 86

Backcross mating after cyclophosphamide exposure, 168
Banana farmers, Israel, DBCP, 40, 41
Birth Defects Monitoring Program, 12, 18–21
Birth defects, time trends, 18–20
Birth weight and prepregnant maternal weight, 49–52
Body weight after paternal cyclophosphamide exposure, 169
Brain tumors, childhood, parental occupational exposure, 141

Cabaryl, 30
Caffeine
   and acetaldehyde and nicotine in rats, 52
   male-transmitted effects, 150
   and maternal risk, 46, 51–52, 54
California
   Los Angeles tumor registry, 137
   vasectomy in, 27
Carbon disulfide, 30, 115, 119
Childbearing, changes in number and timing, and surveillance, 15
Childhood tumors and parental occupational exposure, 137–147
   aircraft industry, 137–138, 140–141
   exposures, summarized, 139
   father's occupational exposure, 144
   information collected from parents, 145
   by job, 146
   Los Angeles tumor registry, 137
   mother's occupational exposure, 143
   occupational histories, 146
   radiation exposure, 138
   sequence of study, 142
   solvent and paint exposure, 137
   tumor type, by occupational information, 140

working conditions and precautions, 138
Clean Air Act, 9
Cliff avoidance, rat, after paternal cyclophosphamide exposure, 152, 153, 158, 160, 168
Confounding variables, 96–97
    age, 94, 96
    multiple exposures, 49–52
    radiation, 97
Congenital Defects Program, Metropolitan Atlanta, 12, 18
Continence time, effect on semen analysis, 36
Control group identification, 94–95
Copulatory behavior, rat, 110–111, 118, 119
Cross-sectional studies, sperm tests, 93
Cyclophosphamide, male-transmitted effects in rat, 151
    active avoidance, 152, 156, 157, 158, 159, 161, 162, 166
    activity, 152, 155, 161, 163
    backcross mating, 168
    body weight, 169
    cliff avoidance, 152, 153, 158, 160, 168
    F_1 cf. F_2 progeny, 169
    paternal exposure protocol, 156, 158–159, 160
    selective phenomena, possible, 168
    spermatogenesis, specific stages, 166
    surface righting, 152
    swimming, 152, 154, 158, 163, 164, 165
    testing battery, 151
    treatment groups, 161
    usefulness of behavioral assessment, 163, 166

DBCP, 11, 16, 25–27, 109
    banana farmers in Israel, 40, 41
    and fertility, 61–62
    pineapple workers, 30
    sperm
        count, 31
        tests, 83, 88, 89
        see also Sperm tests
    summary of studies, 27, 29
    testicular function, 26, 30
    uses and effects listed, 28
Decision-making framework and risk assessment, 2–10
DES, 11, 16
Developmental deficits. See Male-transmitted developmental and neurobehavioral deficits
Dibromochloropropane. See DBCP
Diet and maternal risk, 47
DNA synthesis, unscheduled, 115, 124, 126
Double F-bodies, 87–88, 98
Down's syndrome, 48
Drugs
    during pregnancy, 47
    recreational, 83, 92
    sperm tests as indicators of germ cell damage, 83, 90, 91, 93–94
    recreational drugs, 83, 92
    see also Cyclophosphamide, male-transmitted effects in rat; specific drugs

Ejaculation, rat, 119, 121
Electrophoretic test system (Malling), mouse mutation model, 130–131
EMS, mouse mutation model, 131
Environmental exposure, sperm tests as indicators of germ cell damage, 83, 89
Environmental Protection Agency, 10
Epidemiology, observational, 16–17
2-Ethoxyethanol, 17, 115, 124, 125
Ethylene dibromide, 30, 37–38
Ethylene oxide, 30
    mouse mutation model, 133
Ethylnitrosurea, mouse mutation model, 130

F-bodies, double, 87–88, 98
Female reproductive toxicity, cf. male, 113
    rat model, 119–120
    risk factors, 46–47
    see also Abortion, spontaneous; Maternal risk factors, multiple and common
Fertility as measure of reproductive toxicology, 59–65
    animal studies, 61, 64
    baseline infertility rates, 59–60

biochemical tests, various hormones, 65
comparison of fertility rates, 61–63
DBCP, 61–62
fecundability, 62–63
measurement of individual fertility parameters, 63–65
ovulation, 64
potential, potency and hazard, 63–64
sperm counts, 64
cf. teratology, 59
Fetal alcohol syndrome, 11, 16
cf. male-transmitted effects, 150
see also Alcohol
Fetal loss, early, monitoring, 20
Folate, 47
FSH, 27, 32
biochemical tests, 65

Glioma, 141

Hamster eggs, zona-free, 77–78
Hastings Center, 1
HGPRT mutants, 10
Hypervitaminosis A, 47

Infertility, 59–60; see also Fertility as measure of reproductive toxicology
Interagency Regulatory Liaison Group, 1
Israel, banana farmers, DBCP, 40, 41

Kepone, 30

Laboratory tests, male reproductive function. See Sperm tests as indicators of germ-cell damage in men
Lead, 30
OSHA standard, 26
LH, 27
biochemical tests, 65
Linkage map, mouse, 129
Los Angeles tumor registry, 137

Male reproductive function, evaluation/tests, 67–80, 109–115
animal studies, 110, 112
copulatory behavior, 110–111

pre-exposure semen evaluation, 113
repeated measures model, 114
serial matings, 115
testicular examination, 111
categories, 68
comparative assessment, table, 79
DBCP, 109
dearth of normal data, 109
endocrine evaluation, 70
cf. female studies, 113
interpretation in absence of clinical infertility, 68–69
rat model, 109, 115–126
copulatory behavior, 118, 119
CS_2, 115, 119
ejaculation, 119, 121
2-ethoxyethanol, 115, 124, 125
females, 119–120
MMS, 115, 124, 126
periodicity of spermatogenic cycle, 118
sperm motility, videotape, 120, 122, 123
TCE, 115, 119
timetable, 116
total sperm number, 120
2,4,6-trichlorophenol, 115
unscheduled DNA synthesis, 115, 124, 126
selection, 67–68
semen evaluation, 70–75, 78
sensitivity to early toxicity, 78
sperm
-cervical mucus interaction, 75–76
-oocyte interaction, 76–78
cf. other species, 67
testes, physical examination, 69–70
vasectomized men, 70
videomicrography, 72–74
zona-free hamster eggs, 77–78
see also Sperm; Sperm tests; Testicular function
Male-transmitted developmental and neurobehavioral deficits, 149-166, 168–169
agents listed, 150
cf. maternal, 149–150
see also Cyclophosphamide, male-transmitted effects
Marijuana, 92, 97, 150

Maternal risk factors, multiple and common, 45–55
  age and parity, 48
  alcohol, 46
  caffeine, 46, 51–52, 54
    and acetaldehyde and nicotine in rats, 52
  confounding effects of multiple exposures, 49–52
  dietary pattern, 47
  drugs during pregnancy, 47
  maternal weight gain, 48
  measurement of common exposures, 53–54
    case-case approach, 53–54
    case-exposure study, 54
    maternal recall, 53
  personal habits, 46–49
  prepregnant maternal weight, perinatal mortality, and infant birth weight, 49–52
  sample size, 54–55
  smoking, 46, 50–52, 55
  synergistic effect of related exposures, 52–53
  Thalidomide, 47
  women cf. men, epidemiology, 46–47
  women in workforce and surveillance, 15
Medulloblastoma, 141
Methadone, 150
Methylmethane sulfonate, 115, 124, 126
Morphine, 150
Mouse. See Mutation testing and evaluation, mouse model
Multiple exposures
  acetaldehyde, nicotine, and caffeine in rats, 52
  confounding effects, 49–52
  synergistic effects, 52–53
Mutagenic effects, germ cells, risk assessment, 10–11
Mutation testing and evaluation, mouse model, 129–133
  dominant lethal test, 131
  electrophoretic test system (Malling), 130, 131
  EMS, 131
  ENU, 130
  ethylene oxide, 133
  genetic background, effect on sensitivity to mutation, 132
  homology of genome with human, 129
  linkage map, 129
  semen-locus visible test system, 130
  sex and germ cell state, 132–133
  short generation time, 129
  susceptibility of different portions of genome to mutation, 131–132

National Ambient Air Quality Standards, 4
National Journal, 2
Neural tube defects
  Appalachia, 13, 18
  folate and zinc, 47
Neurobehavioral deficits. See Male-transmitted developmental and neurobehavioral deficits
Nicotine, acetaldehyde, and caffeine in rats, 52
NIOSH, 17, 20

Observational epidemiology, 11, 16–17
Occupational exposure
  and risk assessment, 8
  women in workforce and surveillance, 15
  see also Childhood tumors and parental occupational exposure; specific agents
Occupational studies, male reproductive effects, 25–43
  cabaryl, 30
  carbon disulfide, 30
  data analysis, 31
  DBCP, 25–27, 40, 41; see also DBCP
  EDB, 30, 37–38
  ethylene oxide, 30
  exclusions from study, 32
  negative study, 31
  OSHA lead standard, 26
  problems, 25, 31
  sperm
    collection, 32–33
    count, 25, 34–38, 40–41, 43
    motility, 32
    see also Sperm tests
  spontaneous abortions, 25–26, 38–39, 41–43

studies in Romania cf. U.S., 26
Y body test, 32
Office of Science and Technology Policy (White House), 1, 2
Oligodendroglioma, 141
OSHA, lead standard, 26
Ovulation as measure of reproductive toxicology, 64
Ozone, 6, 7

Paint exposure and childhood tumors, 137; *see also* Childhood tumors and parental occupational exposure
Parity and maternal risk, 48
Peer review, 93
Pineapple workers and DBCP, 30
Population identification, 88, 93
Potency, relative, and risk assessment, 5
Prolactin, biochemical tests, 65

Radiation exposure
    and childhood tumors, 138; *see also* Childhood tumors and parental occupational exposure
    confounding variable, sperm tests, 97
Rat, acetaldehyde, nicotine, and caffeine in, 52; *see also* Cyclophosphamide, male-transmitted effects in rat; Male reproductive function, evaluation/tests
Recall, maternal, risk evaluation, 53
Recreational drugs, sperm tests as indicators of germ cell damage, 83, 92; *see also* Drugs; specific drugs
Regulatory decisions and risk assessment, 3, 9–10
Risk assessment and evaluation
    for adverse reproductive outcomes, 11–13
    end-points and toxicological studies, 12
    observational report, 11
    animal studies, 4, 6
    characterization, 3, 5–8
    decision-making framework, 2–10
    exposure characterization, 7

future directions, 13
germ cell mutagenic effects, 10–11
identification, 3–5
major health end-points, 3, 12
occupational exposures, 8
potentially hazardous chemicals, 3
quantitative risk-assessment, 9
and regulatory decisions, 3, 9–10
relative potency, 5
susceptibility, differences in, 7, 8
see also Maternal risk factors, multiple and common
Romania, occupational studies, cf. U.S., 26

Sample size, 54–55
Semen evaluation. *See* Sperm tests
Sex differences, reproductive toxicity, 113
    rat model, 119–120
    risk factors, 46–47
SHBG, biochemical tests, 65
Smoking
    and maternal risk, 46, 50–52, 55
    sperm tests as indicators of germ cell damage, 92, 97
Solvent exposure and childhood tumors, 137; *see also* Childhood tumors and parental occupational exposure
Sperm
    -cervical mucus interaction, 75–76
    -oocyte interaction, 76–78
        zona-free hamster eggs, 77–78
    pre-exposure, 113
    see also Testicular function
Spermatogenesis, rat
    cycle, periodicity, 118
    and paternal cyclophosphamide exposure, 166
Sperm tests, 83–101
    advantages and disadvantages, 100–101
    in animals, rule, 100
    automated image analysis, 86
    collection, 32–33, 97–98
    continence time, 36
    DBCP, 31, 83, 88, 89
    double F-bodies, 87–88, 98
    embryonic failure and spontaneous abortion, 99
    environmental exposures, 83, 89

experimental or therapeutic drugs, 83, 90, 91, 93–94
guidelines for study planning, 88, 93–98
   assignment to dosage groups, 95
   collecting and analyzing semen samples, 97–98
   collecting questionnaire data, 95–97
   confounding variables, 96–97
   cross-sectional studies, 93
   gaining access, 94
   identifying control group, 94–95
   identifying populations at risk, 88, 93
   peer-review, 93
   statistics, 98
human monitoring, 88
methods, 85–88
morphology, 38, 73–75, 86–87, 93, 98
motility, 32, 85–86
cf. other measures of testicular function, 83–84
recreational drugs, 83, 92, 97; see also specific drugs
reproductive implications of induced sperm changes, 98–99
sperm count, 25, 71–72, 78, 85
   cell/cc vs. total counts, 43
   data from fertile men, 35
   DBCP, 31
   interpretation, 34–36, 40–41
   as measure of reproductive toxicology, 64
   median and mean, 37
   rat model, 120
   semen analysis/evaluation, 25, 70–75
   and surveillance, 41
statistics, 98
videomicrography, 72–74
see also Male reproductive function, evaluation/tests; Testicular function
Surface righting, rat, after paternal cyclophosphamide exposure, 152
Surveillance, role in monitoring reproductive health, 15–22
   birth defects, time trends, 18–20
   changes in number and timing of childbearing, 15
   defining reproductive hazards, 16–21
      monitoring early fetal loss, 20
      routine monitoring, 20
   detection methods, 16, 21
      animal toxicology, 16–18
      observational epidemiology, 16–17
   problems, 21
   sperm counts, 41; see also Sperm tests
   women in workforce, 15
Susceptibility, differences in, 7, 8
Swimming, rat, after paternal cyclophosphamide exposure, 152, 154, 158, 163–165
Synergistic effects, multiple exposures, 52–53

TCE, 115, 119
Testicular examination, 70, 111
Testicular function
   and alcohol, 46
   and DBCP, 26, 30
   sperm tests cf. other measures, 83–84
   see also Male reproductive function, evaluation/tests; Sperm; Sperm tests
Testosterone, biochemical tests, 65
Thalidomide, 11, 16, 47
Tobacco. See Smoking
Toluene diamine, 30
Toxicological studies
   animal, 16–18
   and risk assessment, 12
2,4,6-Trichlorophenol, 115
Tumor registry, Los Angeles, 137

United States, occupational studies, cf. Romania, 26; see also specific government agencies
Unscheduled DNA synthesis, 115, 124, 126

Vasectomy, 27
   physical examination of testes, 70
VDTs, 16–17
Videomicrography, semen, 72–74
Vinyl chloride, 18–20
Vitamin A toxicity, 47